FALL RIVER

WOMEN WRITERS IN ENGLISH
1350–1850

GENERAL EDITOR
Susanne Woods

MANAGING EDITOR
Elaine Brennan

EDITORS
Patricia Caldwell
Stuart Curran
Margaret J. M. Ezell
Elizabeth H. Hageman
Elizabeth D. Kirk

WOMEN WRITERS PROJECT
Brown University

Fall River

AN AUTHENTIC NARRATIVE

Catharine Williams

EDITED BY

Patricia Caldwell

New York   Oxford

OXFORD UNIVERSITY PRESS

1993

Oxford University Press

Oxford   New York   Toronto
Delhi   Bombay   Calcutta   Madras   Karachi
Kuala Lumpur   Singapore   Hong Kong   Tokyo
Nairobi   Dar es Salaam   Cape Town
Melbourne   Auckland   Madrid
and associated companies in
Berlin   Ibadan

Published by Oxford University Press, Inc.,
200 Madison Avenue, New York, New York 10016

Library of Congress Cataloging-in-Publication Data

Williams, C. R. (Catharine Read), 1790–1872.
Fall River, an authentic narrative / Catharine Williams:
edited by Patricia Caldwell.
p.   cm.   -- (Women writers in English 1350–1850)
I. Caldwell, Patricia.   II. Title.   III. Series.
PS3319.W587F34   1993   813'.3--dc20   92-16762
ISBN 0-19-508036-X (cloth)
ISBN-0-19-508359-8 (paper)

This volume was supported in part by the National Endowment
for the Humanities, an independent federal agency.

Printing (last digit):
9   8   7   6   5   4   3   2   1

Printed in the United States of America
on acid-free paper

FOREWORD

Women Writers in English 1350–1850 presents texts of cultural and literary interest in the English-speaking tradition, often for the first time since their original publication. Most of the writers represented in the series were well known and highly regarded until the professionalization of English studies in the later nineteenth century coincided with their excision from canonical status and from the majority of literary histories.

The purpose of this series is to make available a wide range of unfamiliar texts by women, thus challenging the common assumption that women wrote little of real value before the Victorian period. While no one can doubt the relative difficulty women experienced in writing for an audience before that time, or indeed have encountered since, this series shows that women nonetheless had been writing from early on and in a variety of genres, that they maintained a clear eye to readers, and that they experimented with an interesting array of literary strategies for claiming their authorial voices. Despite the tendency to treat the powerful fictions of Virginia Woolf's *A Room of One's Own* (1928) as if they were fact, we now know, against her suggestion to the contrary, that there were many "Judith Shakespeares," and that not all of them died lamentable deaths before fulfilling their literary ambitions.

This series is unique in at least two ways. It offers, for the first time, concrete evidence of a rich and lively heritage of women writing in English before the mid-nineteenth century, and it is based on one of the most sophisticated and forward-looking electronic resources in the world: the Brown University Women Writers Project textbase (full text database) of works by early women writers. The Brown University Women Writers Project (WWP) was established in 1988 with a grant from the National Endowment for the Humanities, which continues to assist in its development.

Women Writers in English 1350–1850 is a print publication project derived from the WWP. It offers lightly-annotated versions based on single good copies or, in some cases, collated versions of texts with more complex editorial histories, normally in their original spelling.

The editions are aimed at a wide audience, from the informed undergraduate through professional students of literature, and they attempt to include the general reader who is interested in exploring a fuller tradition of early texts in English than has been available through the almost exclusively male canonical tradition.

SUSANNE WOODS
General Editor

ACKNOWLEDGMENTS

Producing a book is always a cooperative project; producing a series of books such as Women Writers in English becomes more of a cooperative crusade. What began as a small specialized research project has developed into an adventure in literature, computing, and publishing far beyond our initial abstract conception.

At Brown University, many administrators have given the Women Writers Project invaluable support, including President Vartan Gregorian, Provost Frank Rothman, Dean of the Faculty Bryan Shepp, and Vice President Brian Hawkins. Vital assistance has come from the English Department, including chairs Walter Davis, Stephen M. Foley, and Elizabeth D. Kirk, and from staff members at Computing and Information Services, particularly Don Wolfe, Allen Renear, and Geoffrey Bilder. Maria Fish has staffed the Women Writers Project office through crises large and small.

At Oxford University Press, Elizabeth Maguire, Claude Conyers, and Ellen Barrie have been patient, helpful, and visionary, all at the same time.

Ashley Cross, Julia Deisler, and Lisa Gim shared with me their excitement about teaching early women writers, as have many other faculty members and graduate students at Brown and elsewhere. My friends and colleagues locally in Brown's Computing in the Humanities Users Group and internationally in the Text Encoding Initiative, especially Michael Sperberg-McQueen, have stimulated my thinking about texts and the electronic uses and encoding of texts. Syd Bauman and Grant Hogarth have sustained me in the computer and publishing aspects of the project.

Students are the life's blood of the Women Writers Project, and all of the students who have passed through our offices are part of this volume. Most particularly, among those who worked on *Fall River* were Sarah Finch Brown, Jennifer Bomze, Julia Flanders, Deborah J. Hirsch, Jennifer Hofer, Carole E. Mah, Eowyn A. Rieke, Leslie Stern, and Susan B. Taylor.

Other students have contributed mightily to the spirit and vitality of the WWP: Carolyn Cannuscio, Lisa Chick, John Fitzgerald, Nina Greenberg, Daniel J. Horn, Mithra Irani, Anthony Lioi, Jason R. Loewith, Elizabeth Soucar, and Elizabeth Weinstock.

My personal thanks go especially to Sarah Finch Brown and Julia Flanders for their extraordinary dedication to the project, their meticulous scholarship, and their support through the final production details.

ELAINE BRENNAN
Managing Editor

ACKNOWLEDGMENTS

I wish to thank Catherine Imbriglio of the Brown University English Department for her invaluable research assistance.

I am grateful to Dr. M. N. Brown, Curator of Manuscripts, and to the Brown University Library for making Catharine Williams's papers available to the Women Writers Project.

The engraving of Sarah Maria Cornell from the *Brief and Impartial Narrative of the Life of Sarah Maria Cornell* (New York, 1833) and the lithograph of Ephraim K. Avery from *The Correct, Full and Impartial Report of the Trial of Rev. Ephraim K. Avery, Before the Supreme Judicial Court of the State of Rhode-Island, at Newport, May 6, 1833, for the Murder of Sarah M. Cornell* (Providence, 1833) appear courtesy of the Harris Collection and the Rider Collection at the John Hay Library of the Brown University Library.

PATRICIA CALDWELL

INTRODUCTION

Catharine Read Arnold Williams began her career as an "Authoress" at a fortunate moment. By the 1820s, readers in the new United States were growing in numbers, and many of them were eager for cultural uplift and identity. Women were developing an important voice as consumers of literature, and they were avid for "female" books, particularly fiction. Moreover, an aspiring writer needed no longer to look across the ocean for her literary models: she could see that America, too, was producing women and men of letters and something like a genuine literary profession. Granted, the few supremely famous and successful American authors were men, notably Washington Irving and James Fenimore Cooper, who were turning what appeared to be a rather thin body of American history into new versions of romance, fable, and myth. Still, there had been during the revolutionary period, and there were in the early republic, serious women writers of repute: bluestockings like Mercy Otis Warren and Judith Sargent Murray, and novelists from Hannah Webster Foster and Susanna Rowson to Lydia Maria Child and Catharine Maria Sedgwick, whose fiction attempted to deal directly, if still somewhat sentimentally, with the actual fabric of American life.

Catharine Williams shared with these writers a moral fervor and a deep interest in the drama of the new nation. But Williams, a Rhode Islander born and bred, chose something of that state's traditional independence—some would say eccentricity—on her literary path as well as on her personal one. She was a late bloomer who began writing at the age of forty, alone and with a baby to support, in order to survive; and, whether because of temperament or circumstance, she practiced the writing profession for only seventeen of her eighty-five years. In the eight books that she produced, she moved among the genres of poetry and biography and fiction and history without, apparently, trying to master any one of them or even, at times, heeding them as separate genres. She seems not to have had any significant literary friendships. Yet if her life and *oeuvre* make few claims to grandeur, in her own way Williams found her subjects, her voice, and a measure of literary success. And she left us at least one work that, a century and a half later, still wields genuine power and even fascination: the "authentic" account of the life and violent death of a young, pregnant textile-mill worker and of a Methodist minister's trial for her murder near the ill-fated town of Fall River, Massachusetts.

Little would be known of Catharine Williams had she not left a record of her own life, written in the 1850s at the request of a literary acquaintance.[1] Born on the last day of the year 1787, she was the off-spring of two Rhode Island families, the Reads and the Arnolds, that had distinguished themselves in the Revolution and in state government. Despite this "antient & honourable" lineage, young Catharine never experienced that education, encouragement, and protection by powerful male relatives that were enjoyed by many other early American women writers from Anne Bradstreet to Harriet Beecher Stowe. Having lost her mother at an early age, she was left by her sea-captain father in the care of two unmarried aunts who "lived in strict retirement"— "Ladies of the old School," she said, by whom she had been "brought up a Nun." At twenty-three, upon the death of one aunt and the sudden marriage of the other, young Catharine was "launched into the world, with as little practical knowledge of it, ... as a child of 5 years old," too lacking in self-esteem and too frightened to pursue actively the literary life she dreamed of. As Williams remembered in her autobiography, it was only after "years of affliction, when the trials of life had driven her into the world, & compelled her to seek from within" her own means of support, that she dared to become a writer.

Williams married late for a woman of her day—well into her thirties—and removed with her husband and mother-in-law to western New York State. The union was brief and unhappy. In her autobiography, she reproached herself for showing "the least judgment in that of any action of [my] life," suggesting that she was one of the "women of Genius" who tend "to form the most incongruous connections & to be

---

1. Unless otherwise identified, the source for all biographical information about Catharine Williams in this introduction is Catharine R. Williams, "Sketch of Her Life," in the John Hay Library at Brown University. All quotations from this manuscript are reproduced by permission of the Brown University Library.

the most unfortunate in their married relations, of any other class of human Beings." Williams wrote wryly that her marriage to a descendant of Roger Williams (whose first name she did not record) began rather ominously when the bishop who was to perform the ceremony showed up wearing funeral robes; it ended after two years, and the birth of a daughter.

Williams brought her child back to Providence under "the pain & mortification that would have killed some Women," eventually obtained a divorce, and never saw her husband again. After a short-lived experiment as a schoolmistress, she began writing in earnest, publishing a volume of poems (some of them dating from her teen years) in 1828. The following year she brought out the story of *Religion at Home*, which went through several editions, and her career was launched. *Tales, National and Revolutionary*, including a lengthy study of the life of her grandfather, Captain Oliver Read, came out in 1830, followed in 1832 by "a great caricature on fashionable Life & manners," *Aristocracy, or The Holbey Family*. In 1833 the present work, *Fall River, An Authentic Narrative*, appeared. After a hiatus of six years, she produced a *Biography of Revolutionary Heroes* (1839); a sympathetic portrayal of *The Neutral French* (1841) whom the British had deported from Nova Scotia during the French and Indian Wars; and *Annals of the Aristocracy... of Rhode Island* (1843–45). For the next twenty-seven years, Williams published no more. According to Sidney S. Rider, the local journalist who solicited her life story and who later wrote a memoir of her, Williams had acquired a good income from her writings, which, combined with an inheritance from one of her aunts, left her comfortable for the rest of her life. She traveled, conversed, wrote her autobiography, and spent her later years living in Johnston, Rhode Island, with her daughter Amey and then in Providence, where she died in 1872.

Williams appears to have been as independent in her manners as she was in her practical affairs. On a tour of Canada, on being told that she had missed seeing the British Governor-General's carriage drawn by six beautiful white horses, she disingenuously exclaimed that she would have admired the sight, "when the President of the whole United States, rides with but two." When "an Englishman of rank" asked her at a public dinner how she could write about American genealogies since "even your Aristocracy here, don't always know, who their grandfathers were," she retorted, "They have the advantage of yours however, ... for they, don't often know who their Fathers were." Describing herself as a person heedless of her own or other people's wearing apparel, "she was obliged to the kind care of the Ladies with whom she boarded, to see that she went into the street, with suitable attire, & properly put on & she met with many jokes, from her negligence in this respect"; once, having forgotten to put on a fresh dress for a social call at a hotel, she was ushered into the kitchen instead of the ladies' parlor. Sidney S. Rider, in a note appended to Williams's autobiography, depicted this "remarkable person" as

> rather short in stature and quite stout, very slovenly in dress, her Bonnet always on one side of her head—wears no hoops—and most always a dirty dress—her small eyes twinkle between two lightish Brown curls on either side of her head—and with her squeaking voice you have a complete picture of Mrs. Williams, she is an inveterate Talker and is always ready to talk to any one as long as they will listen—she is deeply interested in politics.

Politics and not fashion was indeed her passion: "She has always been a warm Politition," she wrote, "& devoutly attached to the Democratic Party." In 1842 she supported Thomas Wilson Dorr's rebellion for universal manhood suffrage in Rhode Island; she claimed to have helped abolish both flogging in the Navy, "that disgrace to the Country & to humanity," and capital punishment in Rhode Island; she solemnly announced to a Roman Catholic official (who was hosting a reception for her at the time) that she could never be a Catholic because the two things she hated were "King Craft, & Priestcraft." In later years, she "was the first person to raise the Buchanan Flag, in that part of the Country," and she remembered seeing ten Presidents, including the first: "Mrs. W had seen Gen Washington & once stood at his knee, when a small child, & said his features were ... indellibly engraven on her mind ... 'he wore the same benign expression, as represented in the Portraits of him.'"

Williams did not list women's rights among her causes, but in a deep sense, she partook of that conviction, shared by other women writers of her time, that linked womanliness with patriotism, and she attributed her own love of history to female influence. In the preface to her *Biography of Revolutionary Heroes*, she wrote:

> We have at this moment, a very distinct recollection when a little child, of becoming deeply interested to inquire into the history of our Independence, from the conversation of an old lady of that period. The circumstance that occasioned it, was from my inquiring of her "what the bells were all tolling for." (It was on the reception of the news of the decapitation of Louis XVI.) She was weeping; but I recollect she took the handkerchief from her face and drawing me towards her, held my little hands in hers, while she answered solemnly—"My dear, it is for the death of the King of France." "And who," said I, looking up with childish wonder, "who was the King of France?" She replied, while the big tears stole quietly down her cheeks, "He was one of America's best friends. He supplied us with money, when we needed it, sent us food and raiment and all other conveniences we were suffering for, and men to fight our battles. He became our friend when we had no other; a friend in need, and now wicked men have cut his head off." And here her tears redoubled. I remember I left her, and went to look into the street to see if the men were crying, but to my surprise, not one appeared different from their usual manner. I saw several females that day and all appeared more or less affected. And though I did not exactly comprehend the services of the French monarch, yet from that time I conceived the highest respect for the feelings and opinions of women on those subjects, and imbibed a feeling of interest in the history of my country, which has never left me from that hour to this. (17–18)

Far from yielding, then, to the "barbarism" that was "quite fashionable among many of the lords of creation, to ridicule every thing like patriotic feeling among women" (17), Williams was ready to use her talent feelingly in public causes. "Feeling" was as important as "opinion": for her, as for Harriet Beecher Stowe in the next generation, the woman writer's task was not only to uplift and enlighten, but to relieve distress: "to comfort the afflicted, succor the persecuted, strengthen the weak, & raise up those who are fallen." Unlike Stowe, however, Williams expended almost none of this feeling to oppose the injustice of slavery or to defend the native Americans who sometimes appeared in her historical works. Some of the causes she espoused may seem peculiarly remote today. Of all her works, she placed the highest value on her history of *The Neutral French; or, The Exiles of Nova Scotia,* describing it as "undoubtedly the best" of her efforts and a heartfelt commitment to make known the sufferings of a people she felt had been unfairly persecuted. It was a source of pride to Williams that even among British readers to the north whose government she had treated so harshly, "the Work was received with enthusiasm, & many of them, sent word that 'they had shed floods of tears over it.'" She was sure that Henry Wadsworth Longfellow had been inspired by the book to write his poem, *Evangeline: A Tale of Acadie* (1847).

To *Fall River,* however, Williams later made only passing reference, describing the town of Fall River as "a manufacturing place of some importance in RI.[2] & the scene of many romantic incidents"—a pale description that fails to reflect the energy with which Williams involved herself in the life story of Sarah Maria Cornell and in the trial of the Reverend Ephraim Kingsbury Avery for her murder. On the face of it, Sarah Cornell seems an unlikely person to have attracted Williams's sympathy. Cornell was a young, unmarried woman who earned her meager living as a weaver, wandering from one employer and one boarding house to another in the early years of the growing textile industry in New England. She was the child of a respectable but broken family, had gotten into some minor trouble as a youngster, and, following her factory work from place to place, had found some solace and stability in Methodism, which, with its camp meetings, itinerant ministry, emotional preaching, and emphasis on lay participation and sociability, had gathered in many of the urban poor and the dislocated who were no longer comfortable in the older New England churches. Cornell's short, hardworking life ended before she was thirty. Her body was found on a cold December morning in 1832, hanged from a farmer's haystack roof in Tiverton, Rhode Island, near the Massachusetts border. At first, the death was assumed to be a suicide, but subsequent discovery that Cornell was pregnant, a review of her movements around the time of her death, and a note found among her belongings, all indicated that she

---

2. At the time, Fall River fell on the border of the two states. Many events in the Sarah Maria Cornell story cross and re-cross the Massachusetts-Rhode Island line.

might have been murdered. The trail of evidence led to Avery, a Methodist minister in Bristol, Rhode Island, whose path had crossed Cornell's at several points and who had once been her spiritual advisor. Although nothing came of an immediate inquest, and charges were dropped in a subsequent hearing in Bristol, Avery finally stood trial from May to June, 1833, in Newport, Rhode Island, for the murder of Sarah Maria Cornell. It was chiefly that trial (although occasionally Williams returns to the earlier proceedings in Bristol), viewed within the web of lives and circumstances surrounding it, that gave rise to this book.

Catharine Williams's social world was, to say the least, a far different one from that of the young textile worker whose story she set out to tell. For all her eccentricities, she was a member of the genteel class and of the established Episcopal Church, educated, sophisticated, and clearly disdainful of what she considered the excesses of evangelical religion. In spite of these differences in class and temperament—or perhaps because of them—the meeting of Cornell and Williams in the pages of *Fall River* is a challenging literary as well as human experience. One of the challenges is that, whereas Williams could label her other works as satire or history or biography, *Fall River* was and is not easy to categorize. Once Williams turned from politics to the even more controversial subjects of sex and religion, she left the arena of conventional forms and put together a compendium of fact and fancy, of trial records, letters, biography, and "docudrama" that fits no traditional scheme. As a pioneer of nineteenth-century investigative reporting, Williams allowed herself to interrupt the narrative line, to digress, to backtrack, and at times, to exhort the reader directly and forcefully—a reader, it should be added, who was likely to be familiar with details of the story that Williams sometimes mentions without much explanation. For these reasons, it is not always easy for today's reader to follow Williams's treatment of the Avery-Cornell case and its judicial proceedings. Indeed, several recurrent elements bear a special weight in the narrative and call for some amplification.[3]

---

3. The most comprehensive source of information about the facts of the case is David Richard Kasserman's *Fall River Outrage* (see Bibliography).

First is the matter of Cornell's troubled relations with the Methodist church, and her quest for help from the Reverend Ephraim K. Avery, then newly arrived as the Methodist minister in Lowell, Massachusetts. For some years before 1830, when she and Avery first met, Cornell had suffered accusations of immoral behavior in a number of communities where she had found factory work. The Slatersville, Rhode Island, Methodist meeting had expelled her for allegedly indecent conduct with two young men; in Providence, the local meetings, mindful of her questionable reputation, rejected her; and there was further strain with churches and employers in Massachusetts—in Dorchester, Millville, Lynn, and Lowell, where she lived on and off from 1828 to 1830. In Lowell, an unhappy courtship brought fresh gossip about Cornell, and hoping to remove from the town, she was able to obtain a certificate of good standing (which could be used for admission to another church) from the Reverend Avery. But shortly afterward, under pressure from an employer who wanted to ensure her good behavior, and evidently wishing to show a repentant spirit, she returned to Avery with a confession involving several love affairs; this time, he demanded his certificate back and had her ejected from the church. Avery's disapproval followed Cornell to other communities, and she, seeking forgiveness and reinstatement in the church, sent a letter to Avery acknowledging further instances of bad behavior. In 1831, she was back in Lowell, making personal confessions to each of her fellow Methodists while seeking their signatures on a new certificate. Whether all of Cornell's self-accusations were true is an unresolved question, but it seems clear that she desperately wanted an official statement from the Lowell Methodists so that she could have church fellowship elsewhere, and that she was anxious for Avery to give back the letter of acknowledgment she had written, which had not produced the desired result of forgiveness and which in fact had never been answered. These are some of the "troubles at Lowell," which, Williams reports, led Cornell to seek her fateful "interview" with Avery at the Thompson, Connecticut, camp meeting in August of 1832.

That interview resulted in Sarah Cornell's pregnancy, according to Williams; yet, wishing to shield her subject—and specifically, her subject's body—from the "indelicate exposure" suffered in court, Williams passed quickly over much of the medical testimony offered at the trial. The physical evidence essentially concerned two things: the actual state of Cornell's pregnancy, and the condition of her "murdered, mangled remains." Although the fact of the pregnancy was certain, a dispute arose about the length of time involved, with Avery's defense lawyers suggesting that Cornell was pregnant before the camp meeting. Hence, among the "revolting particulars" offered in court—the results of two separate autopsies—were the precise measurements of the fetus, the defense claiming that its eight-inch length was too large for the three-and-a-half-month pregnancy that would have elapsed after the August meeting. There was also a detailed exploration of Cornell's recent menstrual history, including testimony about her bedding and underclothing as proof of her menses just prior to the camp meeting, and a rebuttal from the defense's physician that periodic bleeding could occur after conception. Even more disturbing to Williams was the description in court of the "cruel butchery" of Cornell's body. What Williams preferred not to spell out too explicitly was that the corpse bore bruises and injuries that seemed to indicate a hasty abortion attempt just before Cornell died. One purpose of all this medical discussion was to throw light on the question of suicide versus murder. The prosecution maintained that, given the corporeal evidence, it was a physical impossibility for Cornell to have killed herself, as the defense claimed. A particular point at issue was the type of knot, known as a clove-hitch, found in the hanging rope, a knot that required two hands, pulling in opposite directions, to tighten it, and that thus seemed an unlikely means of self-strangulation. Still the arguments continued: whether the knot was a matter of common knowledge, whether Avery on the one hand or Cornell on the other knew how to tie such a knot, and what could be determined from the marks found on the victim's neck.

Williams's obvious indignation with such proceedings, and with a long parade of witnesses ready to testify to just about anything, colors her presentation of the facts, but at the same time, it energizes the search for information and sets in relief the conflicting testimonies and points of view, placing readers in the midst of the trial and demanding that they remember and make sense of the many bits of information and the numerous voices heard. Similarly, in tracing Avery's movements both prior to and during the day of Cornell's death, the reader is required to follow the geography of Rhode Island in some detail. One such trip involves the question whether Avery could have sent the three letters found in Sarah Maria Cornell's trunk, letters written respectively on yellow, pink, and white paper, in disguised handwriting, responding to Cornell's troubles and proposing a secret meeting with her. Was Avery the person who posted the yellow letter at Warren, the man who handed the pink letter to the steamboat captain at Providence, and the one who bought the white paper and the purple sealing wafer in the shop at Fall River? Considering how the Methodist church kept its preachers on the road or, in the case of Rhode Island, on the water, a good deal of the time, it was not unlikely. Again, reviewing the day of Sarah Maria Cornell's death, Williams takes her reader to Bristol ferry, "the island," the stone bridge at Howland's ferry, Tiverton, and again to Fall River. "The island," Williams's readers would have known, was Aquidneck, the largest of several islands that partly comprise the state of Rhode Island, containing the city of Newport at its southern end, and at its northern tip separated by water from Bristol on the western mainland and from Tiverton on the eastern mainland. A person who wanted to travel from Bristol, the town where Avery was then living, to Fall River, where Sarah Cornell lived, could do so by traversing this island. On the day of Cornell's death, Thursday, December 20, 1832, Ephraim Avery rode the Bristol ferry to Aquidneck, allegedly to see about buying some coal; but his activities between arriving on the island in the early afternoon and his reappearance at the ferryman's house around nine o'clock at night were the subject of much wrangling at the trial. The prosecution held that Avery made his way across the island to its eastern shore, walked across Howland's ferry stone bridge to Tiverton on the other side, and then turned northward to Richard Durfee's farm outside of Fall River, where, in the early evening, he held a prearranged meeting with Sarah Maria Cornell, and murdered her. Avery's defense contended that he had spent the entire day wandering about the island, but, as Williams demonstrates, there were conflicting testimonies by islanders who were supposed to have seen the minister during the day, and several attempts were made to discredit the witnesses who contradicted his claims.

Here again, the litany of names and jumble of voices can be confusing, but they also convey a sense of the turmoil and complexity of these events and of an underlying emotional urgency that seems to surpass even the sensational facts of the case to tell something about the historical moment. For this reason, and from a profusion of such detail, seasoned by her own forthright commentary, Williams succeeded in fashioning an "authentic narrative" not only of the Avery-Cornell case, but of the world she lived in. Frankly playing a part in her own tale, she made no attempt to conceal her feeling for the "poor factory girl" who, she felt, had no one but Williams to speak for her. The story of *Fall River* will satisfy no one's desire for an objective, balanced report; instead, it brings to life a young and ordinary woman's experience—of hard labor, dislocation, loneliness, violence, and injustice—that was as real a part of "happy America" in the Jacksonian years as industry and expansionism and democracy.

Selected Bibliography

Clarke, Mary Carr. *Sarah Maria Cornell, or The Fall River Murder: A Domestic Drama, in Three Acts.* New York, 1833.

Cott, Nancy F. *The Bonds of Womanhood: "Woman's Sphere" in New England, 1780–1835.* New Haven: Yale University Press, 1977.

Kasserman, David Richard. *Fall River Outrage: Life, Murder, and Justice in Early Industrial New England.* Philadelphia: University of Pennsylvania Press, 1986.

Kelley, Mary. *Private Woman, Public Stage: Literary Domesticity in Nineteenth Century America.* New York: Oxford University Press, 1984.

Kessler-Harris, Alice. "Forming the Female Wage Labor Force: Colonial America to the Civil War." *Out to Work: A History of Wage-earning Women in the United States.* New York: Oxford University Press, 1982. 3–72.

McLoughlin, William G. "The Second Great Awakening, 1800–1830." *Revivals, Awakenings, and Reform: An Essay on Religion and Social Change in America, 1607–1977.* Chicago: University of Chicago Press, 1978. 98–140.

————. "Untangling the Tiverton Tragedy: The Social Meaning of the Terrible Haystack Murder of 1833." *Journal of American Culture* 7.4 (1984): 75–84.

Paul, Raymond. *The Tragedy at Tiverton: An Historical Novel of Murder.* New York: Viking Press, 1984.

Williams, Catharine R. "Sketch of Her Life." The Rider Collection. John Hay Library of the Brown University Library, Providence, Rhode Island.

Note on the Text

This text has been transcribed from the first edition of 1833, printed by Cranston & Hammond, Providence, for Lilly, Wait & Co., Boston, and Marshall Brown & Co., Providence. Obvious printers' errors have been silently emended, but nineteenth-century variants in punctuation and spelling have been retained. Where necessary for clarity, quotation marks have been regularized to conform with modern usage. The presentation of letters (correspondence) within the text has been modified for consistency. The author's notes are designated by asterisks or daggers; numbered footnotes are by the editor.

Fall River,
An Authentic Narrative

Oh for a Lodge in some vast wilderness,
Some boundless contiguity of shade,
Where rumour of oppression and deceit
Can never reach me more.
My soul is sick with every day's report
Of the world's baseness.

PREFACE

It is with feelings of embarrassment never felt on any former occasion, that the writer of this little volume lays it before the public. The tale which forms the principal part of its contents has been hitherto treated in such an indecent manner, that this, of itself, was nearly sufficient to terrify any one at the undertaking; and it was not until after long and reiterated persuasion, that the author was induced to attempt it. Who first proposed it, is of no consequence: it is sufficient that a very great part of the subscribers and patrons of former works have seconded the request; and if the volume answers no other purpose except proving the wish to oblige, it will certainly answer an important one. But we confidently hope it may answer other and more useful ones.

The History of Fall River, a place which is becoming of so much importance in the manufacturing world, cannot but be acceptable to the public. The anecdotes connected with its revolutionary history are worthy to be preserved. And a fair and candid statement of facts, connected with the late unhappy affair in that quarter, is desirable. As to the trial, it does not treat of things in their proper order, nor cannot: and in the next place, none but what is called legal evidence is admissible; and lastly—and its greatest objection—it is not fit for any body to read. A narrative, therefore, that would embrace the facts, without any of the odious details in the trial, is highly necessary, if public curiosity on the subject is lawful: and who shall say that it is not?

There is another way too, in which it is hoped and presumed this work may prove useful:—as a salutary and timely warning to young women in the same situation in life, in which the ill fated girl was placed, who is the subject of this narrative. On many accounts it may benefit. That baneful disposition to rove, to keep moving from place to place, which has been the ruin of so many, will here receive a check. And what is more important still—though an extremely difficult subject to treat upon so as to be understood—they will be warned, by the fate of one, against that idolatrous regard for ministers, for preachers of the gospel, which at the present day is a scandal to the cause of christianity; which neither honors God or benefits his church; and certainly is calculated to bring reproach and ridicule on the christian character. To venerate the ambassador of the Most High, and listen to him with respect, while in the sacred discharge of his ministerial duties, is right and proper; to contribute to his relief in sickness and support in health, of our abundance, or our personal exertion, if necessary, is likewise our duty; but here let us stop, and not make ourselves, and the cause we profess to be engaged in, ridiculous, by such attentions as mortal man *ought never* to receive.

The absurd custom of crowding round some handsome preacher on every occasion, in order to share his smiles, and be distinguished by his *gracious* gallantries, has justly excited the ridicule of a large part of the community, and armed every scoffer with weapons against that holy cause, which ought not to suffer from the faults of its ignorant professors, but which they nevertheless confound together. Besides, ministers are mortal men; and, with good intentions, sometimes persons of weak minds: and it requires a very strong mind to resist continual flattery. Some of them too are ignorant persons; people, who, if they had their proper places in society, would be hewers of wood and drawers of water, rather than teachers. This description of false teachers is very plainly set forth in the Scriptures, as being "ever learning, and never able to come to the knowledge of the truth;"[1] as "creeping into houses, and leading captive silly women,"[2] &c. &c. This last description of preachers take care to insinuate themselves into every place where they can possibly

---

1. 2 Tim. 3:7. 2. 2 Tim. 3:6.

find entrance. No matter what the religious privileges of the people may be, they go among; unless they themselves have built up a sect among them, they consider them as destitute of truth and the means of grace. If a neighborhood is furnished with ever so many good, respectable, competent teachers, supported by those who are able to do it, there must be one more added, if there is no other way to support him but out of the hard earnings of the poor. Now the fact is, that a preacher, who cannot be supported without drawing upon the charity of poor factory girls, ought to go in and go to work himself.

It will be seen too after perusing the history of this unfortunate girl, whether a course of spiritual dissipation is favorable to the growth of religion in the soul; whether a continual round of going to meetings night and day, is in reality recommending the cause, or likely to recommend the character, or preserve the characters of young women, in an especial manner. It is much to be feared it is otherwise. In the first place, this appearance of superior devotedness, this over zeal, fails in no instance to draw all eyes upon her. There is rivalship in churches it is known, as well as in other communities, and such members are watched with jealous regard; if they go and return protected as they ought to be by one of the other sex, barbarous insinuations will sometimes be made; if on the contrary they wander about from meeting to meeting alone, they are immediately censured. And added to this it is expected that the general deportment of such females should differ from that of others; that it should present an appearance of stiffness and restraint incompatible with youth, with cheerfulness, and a social temper; hence the slightest deviation from the prescribed forms is censured in such persons as a crime; what would pass in others without remark, is the subject of unqualified abuse in these, and induces a species of persecution, that too often results in loss of character to the victim.

And is this counterbalanced by any inward advantage? Does religion thrive most in noise and tumult? Does the heart become better, the imagination purer, the temper more placid? can that God, who is worshipped only in spirit and in truth, be only honored in a crowd? Let every heart decide the question.

With respect to embellishment in this book, no person acquainted with the facts, who has seen it, pretends to say there is any, except in the first interview between the physician and the unfortunate heroine of the tale; where it is said the phraseology is improved without altering the facts. If the error is on the side of delicacy we hope to be pardoned.

CHAPTER I

Situated on a rather abrupt elevation of land rising from the northeast side of Mount Hope bay, distant about eighteen miles from Newport, and nine from Bristol, R. I. stands the beautiful and flourishing village of Fall River, so called from the river, which, taking its rise about four miles east, runs through the place, and after many a fantastic turn, is hurried to the bay over beds of rocks, where, before the scene was marred by the hand of cultivation and improvement, it formed several beautiful cascades and had a fine and imposing effect. The village is now only picturesque from the variety of delightful landscape by which it is surrounded, the back ground presenting a variety in rural scenery—where neat farms and fertile fields shew themselves here and there, between hill and dale and rock and wood. The soil, though for the most part fertile, is in some places exceedingly rocky, and often in the midst of such places some little verdant spot shews itself, looking, as Cunningham says, "as though it were wrested from the hand of nature."

But Fall River is chiefly inviting as a place of residence from the salubrity of its air, and the vicinity of Mount Hope bay, which spreads before it like a mirror, and extends easterly until it meets the waters of Taunton river, forming on each side numerous little creeks and coves, which add to the charms of the landscape materially; while on the southwest it takes a bold sweep, and passing round through Howland's ferry, where it is compressed through the narrow channel of a drawbridge, having the island of Rhode-Island on one hand and the town of Tiverton on the other, again expands and flows on to meet the ocean. Howland's ferry is not visible from the village of Fall River, though it is from the bay when at the distance of three or four miles. Vessels do sometimes pass and repass through the drawbridge at Howland's ferry to and from Fall River and Taunton; but the most usual way of access to the former is through Bristol ferry, two miles south of Bristol port. It requires no great effort of imagination to go back a few years, and imagine the Indian with his light canoe sailing about in these waters, or dodging about among the rocks and trees. The neighborhood of Fall River has been the scene of frequent skirmishes among the Picknets, the tribe of King Philip, and the Pequods and Narragansetts. Uncas too, with the last of the Mohicans and the best, has set his princely foot upon its strand.

Fall River, which in 1812 contained less than one hundred inhabitants, owes its growth and importance principally, indeed almost wholly, to its manufacturing establishments; which, though not splendid in appearance, are very numerous and employ several thousand persons collected from different parts of the country, as well as many foreigners: the immense fall of water here being now nearly covered by establishments of various kinds.

There are at least forty thousand spindles in operation, and it is only twenty-one years since the erection of the first cotton manufactory. Previous to this the land in this vicinity belonged principally to the families of Borden, Bowen, and Durfee; three families from whom the principal part of the stationary inhabitants sprung. The land now divided among the different manufacturing establishments is principally held in shares, that is in the neighborhood of the establishments. So flourishing has business been there, that there is scarce a mechanic, trader, or even labourer, who has been there for any length of time, who has not acquired an estate of his own. In 1812 the first cotton manufactory was erected by a company incorporated by the name of the Fall River Company. In the same year, another company was incorporated called the Troy Manufacturing Company, and another factory built. There are now, in 1833, thirteen manufactories, viz. two cotton manufactories of the Troy Company—Pocasset, one woollen co.—New Pocasset—Massasoit—Olney's mill—Calico works—Fall River Company's mills, three in number—Annawan—Iron Works and Nail Manufactory. The Calico Works alone, which cover a large area of ground, employ nearly three hundred hands; its state of improvement is not, we believe, exceeded by any establishment of the kind in the country—besides a number of machine shops, &c. which, stuck about on the jutting rocks, many of them in the very bed of the stream, have a most singular appearance. The fall originally was through a deep black gulf, with high rocky sides. Across this gulf most of the manufactories are built. There is an appearance of active industry and a spirit of enterprise, as well as of cheerfulness and contentment, that at once strikes a stranger. It is evident too from the number of houses of worship, schools, &c. that the moral and religious education of the rising generation is not neglected. There are seven houses of worship. Two for Congregationalists, two for Baptists, one Free-Will Baptist, one Unitarian, one Methodist &c. There are a number of free-schools here, towards which the inhabitants themselves voluntarily contribute twenty-five hundred dollars per annum. The number of inhabitants at the present date, 1833, is said to exceed five thousand. It is to be supposed that among the heterogeneous materials which form the community in this place, there is a great variety of character, as well as of creeds; occasionally some differences of opinion as well as clashing of interests. Yet for the most part crime has been unknown there. There have indeed been a few suicides, but they were "few and far between;" and it has often been a boast among the inhabitants, that living as they do, on the borders of two states (part, and by far the greater part, is in Troy, Mass. the other in Tiverton, R.I.) the laws of either were seldom called in to punish any thing except venial transgressions. Fall River too can boast of its prowess in battle, of its revolutionary characters in "the times that tried men's souls."[1] For although their

---

1. "These are the times that try men's souls" opened the first of Thomas Paine's sixteen "Crisis" essays, written from 1776 to 1783 and collectively titled *The American Crisis*.

humble attempts to resist invasion have not yet found a place on the pages of history, yet certain it is, the tide of war has once rolled its threatening waves as far up as to reach the shores of Mount Hope bay. The character for bravery, generosity, and independence of mind manifested at that period seems to have become a part of their inheritance. Among all the changes which the increase of population causes, the primitive virtues of simplicity and hospitality are still eminently conspicuous. Whoever goes to reside there seems to adopt readily the manners of the inhabitants. Even the labouring part of the community in the manufactories, as well as in other departments, are positively distinguished by a degree of refinement and courtesy of manners, superior in a great degree to what we usually meet with in manufacturing villages.* It is a fact that speaks loudly and deserves to be recorded in letters of gold, that Fall River is the only place known to the writer of these sheets where she ever past a week without hearing one individual speak ill of another: and few persons ever had a greater opportunity, having spoken with more than three hundred people during that time.

We have stated that previous to the commencement of its settlement as a manufacturing village, and even as far back as the revolutionary war, the families of Borden, Bowen and Durfee, were the principal proprietors of the soil—and brave fellows they were too, some of them. Even the soil around this secluded spot was stained with the contest. At the time Newport was in possession of the British, there was an attempt made to destroy their mills at this place, consisting of saw mills, grist mills and a fulling mill. An expedition was fitted out in boats, and came upon them in the night with the

---

* I shall always recollect with pleasure one little incident, in one of the weaving rooms of the manufactory, where the noise was very distracting arising from a vast number of looms going at once. The machinery suddenly stopped, and a strain of music arose simultaneously from every part of the room, in such perfect concord that I at first thought it a chime of bells. My conductor smiled when I asked him if it was not, and pointed to the girls, who each kept their station until they had sung the tune through.

intention of firing the village, consisting of a little cluster of houses, about ten in number, and those remote from each other. They were aided by some of the tory refugees, and succeeded in landing on the shore, a little below the long wharf, that now is, where they fired the house of Thomas Borden. Several little bridges lay between them and the mills, and these were immediately destroyed by the brave little handful of men collected on the spot, except the last, behind which they entrenched themselves, and commenced firing a few yankee shot, and from behind the house of Richard Borden, at the corner of which one of the enemy was shot. (The old fabric is still standing.) The enemy continuing to advance, and becoming more formidable, they succeeded in levelling two of them; one was shot dead, supposed by Doct. John Turner, and the other mortally wounded. This rather intimidated the assailants, who made a motion to retreat, but after halting at a little distance, returned again, and the scuffle was renewed—the yankees fighting bravely, with their last powder and ball; finding their ammunition all expended, they contrived to make up the defect by management. One of them, Sherman, by name, was mounted on the wall, and instructed to give orders, which he did with a great flourish, telling them to fight on bravely—the day was their own, and they had ammunition enough to last a month. The poor fellows had then the very last in their guns—but they gave a great shout, and discharged that in the face of the foe, who swallowed the bait, and retreated to their boats, carrying with them, however, one prisoner, old Mr. Richard Borden, who had ventured too near in the zeal of the moment. Boys fifteen and sixteen years old, fought in that contest, and women brandished their broomsticks—and tradition says only one small boy was frightened, and he ran off and hid in the woods until it was over. One of the tories who had been an inhabitant of Fall River, and guided the enemy to this little nook was named Holland. The business not prospering as he expected, he was glad to retreat with the British, and at the evacuation of Newport went to reside at Halifax. Many years subsequent to this, and after he had become quite an old man, he returned to America, and being anxious to see Fall River, the scene of his treacherous attempts, he visited it under an assumed name. Thomas Borden was then an old man, and the stranger made some pretence for calling at his house, but in spite of his disguise and the lapse of years, his eagle eye detected the resemblance, and hastily advancing he demanded to know "if he was not the traitor Holland." The stranger stoutly denied himself to be that character. "If I knew you was," said the old man, clenching his fist, "I would lay you on that forestick, (pointing to the fire) and roast you to a cinder." Holland, terrified, fled again from the place and has never been there since.

This Richard Borden was a singular character for oddity. He was taken prisoner, I before observed, at the memorable contest of the mills, and as they were carrying him off laid down in the boat while they were passing Bristol Ferry, lest some shot from his enraged countrymen should reach them. The enemy commanded him to stand up, which he refusing, two men took hold of him and attempted to force him upon his feet, when a chain shot from the shore, mowed them both down at once, and they fell on the body of the prisoner, dead men.

The wounded prisoner, meanwhile at Fall River, died the next day, and the two comrades were buried on the spot where they fell, side by side. One Peter Thatcher, who had distinguished himself on that memorable night, advanced to the grave while this operation was performing, and protesting if their heads were laid together there would be some mischief hatching, commanded them to be laid heads and points. This was accordingly done, and in 1828, when the ground was excavating for the erection of the Massasoit Factory, the bones of the unfortunate victims of kingly power, of the poor wretches dragged from their families three thousand miles across the water to engage in a broil of which they probably knew nothing— were discovered laying heads and points.

> War is a game, that were their subjects wise
> Kings could not play at.[2]

The growth of Fall River from the period of the revolution to the year 1812, must have been slow—and even since that, until 1822,* when there was but four stores in the place, of any description, and not to exceed four hundred inhabitants. There is now about 100 shops and stores of various descriptions—but excepting two or three on the Tiverton side of the village, scarce any where spirituous liquors are retailed, and not a single distillery in the place.

The roads north and south of the village, lead through a delightful country. The view of the island of Rhode-Island on the south one is beautiful, almost enchanting—while that leading to Taunton is scarcely less picturesque. On this road lies the little village of Assonet, where there is considerable commerce carried on. It is a singular sight to see large vessels coming up to the very doors of the cottages, sheltered and shut in by the little woody point that encloses the tiny harbor—and music to hear the voices and loud laugh of numerous little urchins who are frequently seen playing on the hull of some old vessel on the grassy strand. These fairy landscapes on the one hand, are strangely contrasted by the wildness and sterility of that on the east, which resembles a newly settled country. The land lying between Fall River and New-Bedford, a distance of from sixteen to eighteen miles, is a perfect desert for the most part, being only diversified by bogs, rocky pastures and forests of scrub oak and wild poplar. The village of Fall River perhaps owes much of its picturesque appearance to the rocks which are seen rising on all sides, some of the most finished buildings being nearly surrounded by rocks.

---

* The third manufactory was erected in 1821, and two more in the ensuing year. At this period, 1833, a large and elegant one is going up.

---

2. William Cowper, *The Task* (1785), bk. 5, lines 187–88. Williams slightly alters the wording.

It almost seems in the law of destiny that every place shall have something in its history to recommend it to the attention of mankind.

In the countries of Europe, in every part of the old world, scarce a village or hamlet is past, where the attention of the traveller is not called to some circumstance of notoriety connected with the history of the place, either it has been the birthplace of some hero, or statesman, or poet, renowned in the annals of the world, or the spot where some bloody battle has been fought that perhaps decided the fate of nations. Here was once the resort of banditti, and here once stood the monastery of some religious fanatics. Here was the cell of an anchorite, and here the home of unbounded luxury and unbridled licentiousness. Those ruins cover the springs once so celebrated in history where the beauty and fashion of centuries long gone by resorted for health and pleasures, and drank from the fountains now hidden fathoms under ground. This place witnessed ages since, the vows of those celebrated lovers, and this was the scene of a black and midnight murder. Here, dwelt the witches of yore, and here the sorcerers. Here was lighted the fires of the martyrs, and there, their persecutors breathed their last. Here wept an injured, banished queen, and here a king abdicated his throne. In short, there is no end to the catalogue of events by which each place is consecrated in the memory of man.

In our happy country, new to crime and unknown to greatness comparatively speaking, there is little of this kind of distinction known. It is sufficient that the thriving city exhibits the appearance of industry and application and enterprise, that the rural landscape teems with sights and sounds of human happiness, that it is clothed with the flowers of spring, the verdure of summer, and the fullness of autumn. The unenvied distinctions conferred by the monuments of former greatness and vengeful crimes, we desire to leave to our older neighbors. Yet even in this our new and favored country, crime is sometimes known. The primeval curse which extends over the whole earth, has not left our plains and vallies without some demonstrations of its universality. The blood of man, slain by his brother man, has at intervals stained the soil where peace and purity were wont to dwell, and the cry of murder, borne on the midnight blast, has sometimes been heard, even in some of the most secluded parts of happy America. The traveller in future ages, as he wends his way through the delightful village we have been describing, shall point to the lowly grave on the side of yonder hill, and say "even here, has the curse been felt—even here, has murder stalked abroad, amidst scenes of nature's loveliness, calculated to warm the coldest heart with gratitude towards that good and glorious Being who clothes the fields in plenty and bids the landscape smile, has the assassin lurked—here plotted the direst deed of darkness—here executed a scheme of cruelty which the savages of our western woods might have shrunk from." Here at this lonely grave, whose plain and unobtrusive stone just tells the *name and age of a female,* cut off in the prime of her days—and tells no more—shall the young and the beautiful read the warning against the wiles of man, here try while recounting the sad story of her who sleeps beneath to fortify each other against the encroachments of vice, especially of that which captivates under the mask of love. Here the prudent mother shall bring her lovely daughters to read those lessons of prudence and caution, which of all other lessons the youthful heart is most apt to revolt at, the youthful mind to forget. And pointing to this place, the drunkard, the swearer, the Sabbath breaker, the gambler, and even the highway robber, shall exclaim, "that grave attests that monsters have lived worse than me!" On yonder slope, where nature has spread her richest carpet of almost perpetual verdure, and where the quiet of the scene might seem to speak of sweet repose and heavenly contemplation, a deed of darkness has been perpetrated, at which even such might have revolted. But this is digression, and we hasten on to the story.

CHAPTER II

About half a mile from the centre of the Village of Fall-River, in a southwardly direction, on the direct road to Howland's ferry and rather remote from any other dwelling, there is a large old fashioned farm house belonging to a family by the name of Durfee. The land descends from here towards the bay with a gentle slope, and is probably about 150 or 200 rods to the water. The house stands in the State of Rhode-Island, and is quite near the line that marks the boundaries of the two states. Proceeding from here towards the village you enter the suburbs of Fall-River. In the State of Massachusetts in about a quarter of a mile distance, within a short distance from this line on the Mass. side is the residence of a physician esteemed in his profession as well as in his private life, of unaffected manners, and unassuming deportment. His appearance is the very epitome of plain old fashioned Republican simplicity: there is a degree of frankness and benevolence expressed in his countenance that at once secures the confidence, even of strangers.

It was on the evening of the 8th of October 1832, that the Doctor was summoned to the parlor to see a lady who desired to speak with him. This circumstance to a physician was nothing extraordinary, and therefore it was without any feelings of curiosity or awakened attention that the doctor obeyed the summons: he perceived a young woman very plainly habited and of most dejected appearance: her age he judged might be about twenty eight, and her countenance bespoke the possession of beauty in happier days—but it was now clouded with care and shaded with grief, and as she arose to address the doctor upon his enterance, the air of extreme dejection that she wore, caught his eye, and in a moment interested him in behalf of the unknown sufferer. He begged her to be seated: while drawing a chair opposite, he endeavoured to penetrate so deep a grief and ascertain the cause of this visit. "She had come she said to consult him on the subject of her health. She had not been well for some time, and wished to ascertain with certainty the nature of her disease."

The doctor desired her to mention her symptoms. She did so. Not having the slightest recollection of seeing her before, he inquired, was she a stranger here? "Not exactly; she had been employed to weave in one of the cotton manufactories for some time past." "Her work probably disagreed with her: had she been used to such employment?" yes, for several years. "Would she be so good as to state her symptoms once more?" she did so, with a faultering voice, and changing cheek. The doctor rose, took a turn or two across the room, and again seating himself opposite, asked the question, "Are you married, madam?"

"No sir," said the young woman faintly.—A long pause ensued.

"If you were a married woman I should be apt to tell you what I thought, but as it is I scarcely know what to say, except it is my opinion you will not be able to work in the factory much longer." The miserable young woman clasped her hands together and wept profusely.

"Can you speak with certainty, sir, as respects my case?" "I cannot," said the doctor, "nor no other person with certainty. I only give my opinion, grounded upon the facts you have stated with respect to your ill health, and I may add your too evident distress bespeaks you have been the prey of a villain; but has not the person who has thus entailed misfortune upon you, the power to take you from the hardships of a factory and place you in a comfortable situation, until you can again resume your employment with safety to yourself?"

"I am afraid he would not be willing to do so."

"Not be willing! then he must be a very base man. It certainly is in his place to do so. Who is he?" no answer but tears.

"Can you not tell me his name?"

"I cannot, I dare not," said she at last, bursting into a fresh flood of tears.

"Have you no connections in this place young woman?" demanded the doctor.

"None sir except religious connections."

"Then you are a member of some religious society—of which?"

"Of the methodist, sir."

"Well your case is certainly a very peculiar and a very distressing one, but I can see no reason why you cannot tell who this person is—this man who has led you into this trouble—there seems to be some great mystery about it, which I am desirous of unravelling. Perhaps I can advise you to some safe course, and if I am to be your physician I insist upon knowing before I give you any further advice, and if it is in my power to befriend you in any way I should certainly do so." It was not until many apparent struggles with herself, much persuasion, solemn injunctions to secresy, and finally a promise on the part of the doctor not to expose the name, that she at length reluctantly disclosed it; and great was the doctor's astonishment indeed when she named a preacher of the gospel as her betrayer—a Methodist minister!

"Monstrous!" said the appalled physician, "and does he preach now?" "Yes sir, in Bristol, next town to this."

"But how, where, which way, could a minister of the gospel contrive to insult one of his flock? Where young woman, I must ask, has your interviews taken place?"

"Our *interview*, sir, was at the late Camp Meeting in Thompson, Con. It was unsought by me for any such purpose, but I trusted myself with him in a lonely place, and he acted a treacherous part."

"Amazing," exclaimed the doctor, "under the mask of religion too! Well young woman it is useless to mourn over what is past and cannot now be mended. Your business must now be to take care of yourself—and there is as I conceive one straight forward course for you to pursue. Such a man deserves to be exposed. It is a duty you owe not only to yourself but to the public to expose the man. It is outrageous that such a man should continue to deceive the public. I would therefore if I were you boldly go forward and expose him to the world, and compel him by law to do me justice. You would certainly be doing society a service to unmask such a person."

"Oh I cannot, I cannot sir, indeed," said the young woman, with a shudder. "I cannot consent to bring such disgrace and trouble upon the church, and upon his innocent family too. He has a worthy woman for a wife, and she and all his innocent children must be disgraced if he is exposed."

"Well, I know not what to advise you, young woman, if you are averse to this course. There is but one other way to obtain redress— and that is by threatening him. You must at all events be provided for before long, and the best way is in case you do not expose him, to threaten to do so unless he settles handsomely with you, and enables you to leave the factory until after the termination of this unhappy affair." To this the young woman assented, and saying she would call again, after writing to him, withdrew.

The image of this afflicted and unhappy person could not momently be erased from the mind of the doctor. The circumstance of itself was calculated to interest, and the sufferer, though not very handsome, was certainly a very interesting person. It was not long though before she called again, and the subject of her second communication was certainly not less interesting than the first. She came now she said to ask advice as a friend. She had recently received a letter from Mr. Avery requesting her to come to Bristol and see him there—that he appointed a time and place, and seemed anxious for the interview. She stated also she had received another letter from Providence, during the four days meeting.[1]

The doctor again advised her to compel Avery to a settlement,

---

1. A gathering of local Methodists with ministers from surrounding districts for four days of intensive prayer meetings and religious services.

and she asked what she had better say to him. He observed that she ought at least to demand three hundred dollars, and he had no doubt Avery would think himself well off to come off so.

"Why," said she, "he is not able to give such a sum. The Methodist ministers are poor—all poor. They are very illy paid for their services, and I doubt his power to make up such a sum, besides I should not dare name so much for fear he would think I had told some one." And she seemed to be in considerable terror at the idea that he should suspect he had been exposed to any one. She then informed the doctor that she had a short interview with him at Fall River, where she met him on the meeting-house steps, and walked away with him, and that he wished her to take a medicine which he recommended, in order to prevent future trouble and expense, and at once obliterate the effects of their connexion. The doctor inquired what it was, and was shocked and surprized to learn it was one of such deadly effect that she would probably have expired on the spot had she taken it. The drug referred to was the oil of tansy, one of the most violent things ever used, and never given except in very small quantities, and under the direction of a physician. Comprehending as he now thought a little of the plot, he advised her against a private interview with Avery, and begged her by no means to go to Bristol and give him the private interview he requested, nor to take any medicine of his prescription, telling her the one recommended would probably have killed her on the spot, if not, it would have utterly destroyed her health for ever.* The girl seemed shocked, but could not seem to believe her betrayer had designs on her life. The doctor observed if she meant to do any thing of that sort she must apply to another physician. She however avowed her determination to take nothing, but bear as she said, the whole shame and disgrace of it herself, "and take care of her child as well as she was able." The doctor commended her in this resolution, and told her it was

---

* Thirty drops, she said he told her to take at once. Four drops is considered a large dose.

his duty to come to her, not hers to go to him, and to have him by all means come to Fall River, and meet him in some suitable place, where they could talk it over and make some settlement with him, that was, in case she had still resolved not to expose him publicly. This she seemed resolved not to do, and spake again of the distress such a disclosure would bring upon his family, and mentioned the agitation the first disclosure of her situation had caused him. He protested to her afterwards that he passed the *most wretched night that night he had ever done, having scarcely closed his eyes.* Much more conversation occurred of the same description, accompanied by many tears, which the doctor observed she always shed when conversing on that subject; and thanking him for his kindness, she withdrew, leaving an impression of pity and admiration upon the mind of the good physician, that one so feelingly alive to sentiments of virtue and propriety should have fallen into such a snare. She had, between these interviews mentioned, called for medicine to take, such as her health required, and the doctor observed he never saw her without her shedding tears and betraying most painful feelings with respect to her situation, although she was calm, and seemed to have resigned herself to the event.

A few weeks only elapsed since the last visit of Miss Cornell, during which the doctor often thought of her, and wondered how she was likely to settle the difficulty with her seducer as he termed him, for so perfectly modest and proper was her deportment that he could on no account harbor an opinion, but that she had been artfully led from the paths of virtue, by one in whom it was perfectly natural she should place the utmost confidence. He looked upon her as one of the most unfortunate of women, but could not despise her as he might have done in other circumstances.

It was on a cold frosty morning, the 21st December, that the doctor observed some people running up the street, apparently in great haste; he stood at the window watching when they should return, to know what the matter was; but no body came back, while another and another party followed close upon the heels of the former. The women appeared to be horror struck as they collected in groups at their doors or in the streets, and many leaving their families just as they were, (it was about breakfast time) and hastily throwing something over them pushed on in the direction of Durfee's farm. Presently some one came running into the doctor's, saying a young woman had just hung herself up at Durfee's. The doctor stopped to ask no more, but catching his hat, ran up to the farm, without however having the least suspicion who it was. Upon gaining a stack yard some fifty rods south of the house, he perceived a female lying on the ground, for they had taken her down. She lay with her cloak, gloves and calash on, and her arms drawn under her cloak.

"Does any one know her?" asked one. "She is well dressed," said another, "I think she must be somebody respectable." "Yes I know her," said the Methodist minister who had arrived on the ground a little previous to the doctor—"she is a respectable young woman, and a member of my church."

Just then the physician reached the yard, and hastily lifting the profusion of dark locks that had fallen entirely over her face, he discovered with grief and astonishment the countenance of his late interesting patient. Horror struck, he endeavored to loosen the cord from her neck; it was nearly half an inch imbedded in the flesh. But alas! there was nothing in the usual remedies to produce resuscitation that would have availed any thing here, for the young woman appeared to have been there all night and was frozen stiff. And is this the end of thy sorrows, poor unfortunate! thought the kind physician, as bending over the hapless victim of unhallowed passion. He gazed upon that altered countenance—altered it was indeed—it was livid pale,—her tongue protruded through her teeth—pushed out her under lip, that was very much swollen as though it had received some hard blow, or been severely bit in anguish, gave a dreadful expression of agony, while a deep indentation on the cheek looked as though that too must have been pressed by some hard substance;

but whatever he thought at that time respecting the means by which she came to her death, he wisely forbore to utter it, and the jury of inquest was summoned in immediately.* In the mean time the respectable farmer in whose premises the deceased was found, after having her carefully conveyed to the house, inquired of the Methodist minister if she had any friends in the place, and if not whether the society of which he said she was a respectable member would not see to the expense of her funeral. That person replied that he did not exactly know their rules in such cases, but he would go and consult them and return soon and inform them. Meanwhile the truth struggled hard in the breast of the doctor. He had felt himself bound to secrecy in case the girl had lived, respecting the name of her betrayer, but her death and the awful manner of it impelled him to reveal what he believed to be the cause. He felt that death had taken off the injunction of secrecy; and stepping after the clergyman, he related the confession of the unhappy girl to him, and what she had said respecting his brother Avery. In what language he expressed himself, or whether he gave way to the feelings of indignation which the knowledge of such a transaction was calculated to awaken, is not known, but the reverend listener was at once roused to defend him, and express his full belief that his brother was perfectly innocent, and finally asserted "that the deceased was a very bad character, and that Avery had told him so, and warned him against her, and that she was not in full communion with the meeting but only received upon probation." Very shortly he returned to the house of Mr. Durfee, and said that "the deceased was a bad character and the meeting would have nothing to do with burying her."

---

* Her countenance was exceedingly distorted, and there was not only an expression of anguish upon it, but one of horror and affright, combined with an angry frown. "That terrible look," said the doctor, "was present with me for months, and often in the dead of night has appeared to my imagination with such force as to awake me, and I can scarcely think of it now without a chill. That look never was seen on the countenance of a person who did not die by violence." He expressed his amazement that among all that was said in Court that circumstance was not attested.

Of course Mr. Durfee's astonishment was very great, having just before heard the Rev. gentleman say "she was a respectable woman and a member of their society." But nothing influenced the honest and benevolent farmer to omit his own duty, and deny the right of burial to the poor unhappy girl whose remains Providence seemed in a peculiar manner to have confided to his care.

"She shall have a burial place in my grounds," said he, "near my family, and as respectable a funeral as any body, and as respectable a clergyman as any other to make the prayer, and every thing that is necessary and decent shall be attended to." And without any fear of contamination, from the neighborhood of one whom the clergyman chose to denominate a vile character, he gave orders to have a grave prepared for her near his own family.

In the mean time a hasty and irregular jury had been selected and sitting upon the question, and after a very superficial observation, and no examination whatever of her person, brought in a verdict of "suicide." The corpse was then delivered into the hands of five or six of some of the most respectable matrons of the village who had volunteered to perform this office of benevolence towards the hapless stranger.

They commenced this work with mournful reflections upon the subject of self-murder, and some expressions of pity towards her whose hard fortune some way or other must have driven her to so rash and daring an act, for that she died otherwise than by her own hand never entered the heads of either of the good women. But what was their astonishment when stripping the body for the purpose of arraying it for the grave they discovered marks of violence about her person. "Oh," said one of the oldest of the ladies who they called aunt Hannah, "what has been done?" The person addressed answered "rash violence." Just above each hip were marks of hands, the bruises of which were very bad, so that the spots of the thumb inwards, and the fingers outside were distinctly visible, they were those of a large hand, for one or two of the women applyed theirs and they were not large enough to cover the marks: one only, the person they called aunt Hannah, found her hand to fit: there were bad bruises on the back, and the knees scratched and stained with grass as though they had been on the ground during some struggle: spots below the knee where the skin was rubbed off and bad bruises on the back; the right arm was bent up and the hand turned back, and it was with much difficulty the females could bring it down, after fomenting it for some time with warm water, and when they succeeded in bending it down it snapt so that they thought it must have been broken: appearance of a blow on the under lip, which was much swollen, and the tongue projected out a little. Still those women said but little, except a few whispers among themselves: in fact the time was too short for much talking. The body was not laid out until past noon on the day she was found, and she was buried at one o'clock on the next day. One most startling circumstance however occurred to arouse the attention and petrify the blood of the spectators.

Mr. Durfee, the farmer who found the deceased, took his wagon (shortly after the verdict of the jury of inquest) and proceeded to the house where she had boarded, after her things, the object of this was to find something suitable for grave cloths, and if possible to ascertain by some letters or something of that kind where the friends of this poor girl, if she had any, were to be found. He found a trunk, locked, and a bandbox of clothes, &c. which he took, and returned about noon. The key of the trunk was found in the pocket of the deceased, in presence of a number of persons. The trunk was examined, and four letters found in the bottom of it. One was addressed to the Rev. Mr. Bidwell of Fall River, her minister, written by herself. The other three were anonymous, but directed on the outside to Sarah M. Connell, Fall River. Near the middle of the bandbox lay a small piece of soiled paper and a lead pencil. Mr. Durfee did not open the little piece of paper or think of its being of any consequence whatever. Two of the women, on rummaging the bandbox late in the afternoon, in hope by some means to discover where to direct a letter to her friends, chanced to observe this very piece of paper, which, though very small, soiled, and looking like waste paper, they unfolded and read. It contained these words—"If I am missing enquire of the Rev. E. K. Avery.　　S. M. C."

There were a great many persons in the house, and constantly going and coming, and although the women talked much about it and shew it to others in the house, one of whom was the wife of the congregational clergyman, invited to make the prayer, yet it was not seen by the master of the house until next morning. His thoughts upon reading it may easily be discerned. The hour of the funeral however drew near and active duties prevented much time from being spent in debate. The resolution however of Mr. Durfee and some others to have the matter investigated, seemed to have been taken. A crowd gathered early to the house, and solemn and appropriate prayer was made by the congregational minister, the Rev. Mr. Fowler, and, followed by a numerous procession, the remains of the unfortunate and mysterious stranger were conveyed to the grave. Providence however had determined that though consigned to the grave it should not be to present repose. A storm was gathering which was destined not only to call forth the dead from her grave, but to shake the society to which she belonged to its centre—a storm whose effects have continued to be felt ever since—a contention which has embittered many former friends against each other, created many heart-burnings, assailed the peace of families, hindered the christian missionary in the exercise of his pious duties, caused the name of Christ to be blasphemed, and in some places almost depopulated churches.

CHAPTER III

Although consigned to her grave, the image of the murdered maid (for murdered he now no longer doubted she was) continued to haunt the pillow of Mr. Durfee, and he arose on the following day determined to investigate the dark mystery which hung over her fate. A circumstance occurred on this morning to materially increase the evidence of the murder of the young woman. A man in the neighborhood, (Thomas Hart,) while walking near the scene of the sad catastrophe, found about thirty rods from the place, in the direction towards Fall River, a piece of a comb, which upon being shewn to the woman where the deceased boarded, was identified as hers. It was also known by the jeweller who had mended it for her a short time before, by the rivetting, which was peculiar. This piece of comb, evidently broken in a struggle, was carried by Mr. Hart to Mr. Durfee. That gentleman took it, and with that and the piece of paper found in the bandbox, waited on the coroner.* The case

---

* The first part of the comb was found some rods from the place while she lay near the stack, after they had taken her down, and the man who found it brought it and laid it on her cloak. They did not then know but she wore a broken piece in her hair, until after its fellow was found. Some way further off, on the lonely path leading round the corner of the wall towards Fall River, she was buried with the first piece in her hair, and when disinterred it was taken out and compared with the remaining piece found, and they fitted, and both parts were then identified. It was singular that the pocket handkerchief of the deceased, found near her wound up in a hard bunch and wet through and through, should have been so little thought of at the time. By soaking it in cold water it would have been ascertained it was wet with saliva, but they did not think of this test at the time, though it was afterwards believed to have been used to stop her mouth by some person who murdered her. Doctor Wilbour remarked that the cloak showed marks of tears, which combined with the discharge from the nose appeared to have been very plentifully shed and ran down on eachside of the cloak. He has even expressed his hope "that they might have been tears of penitence as well as anguish, shed when she found the fangs of the murderer were upon her, and she was about to appear in the presence of her God."

seemed to call loudly for examination, and the coroner ordered the body disinterred on the following day, and called a new jury. Three of the principal physicians and surgeons of the place examined the person of the deceased, that is the external bruises, and ascertained she had told no falsehood with respect to her situation. From the state of the lungs it appeared she died of suffocation, and from the mark of the rope around her neck, that she could not have died by hanging, but by the drawing of the cord, which had been drawn so tight as to strangle, and must have been so before suspension from the stake, as the knot they all deposed was not a slip knot, but what is called a clove hitch, and could not have been drawn but by pulling the two ends separately. Various other circumstances now for the first time detailed, were related, such as the deceased being found with her cloak hooked down before and her hands under it, her knees within four inches of the ground, and her clothes smooth under them, and moreover as it was known that when the neck is not broken by hanging, and hers was not, there is a great struggle in death, and there was not on the ground beneath the least signs of any. On the contrary, her feet were quite close together, her clothes standing off from her behind as far as they would reach, and smooth under her. And lastly, and most extraordinary of all, her gloves on her hands, without any marks of a rope or any thing of the kind upon them, although the rope must have been drawn with great strength by two hands before it was tied to the stake.

With all these proofs before them it was not surprising their verdict should be "murder." It was true suspicion pointed at Avery before, but the supposed sanctity of his character shut the mouths of many who but for that and his profession would have been ready to exclaim "thou art the man."[1]

Although Mr. Durfee and others were thus alone in acting, it must not be supposed that circumstances of the nature just

---

1. 2 Sam. 12:7.

described could be concealed. They were not: and the inhabitants of Fall River on the Massachusetts side (where they do business off hand, and not quite so clumsily as in Rhode-Island) having heard from the first the circumstances of suspicion that had been developed, became very much amazed at the slowness of enquiry respecting such a horrible transaction; and feeling themselves rather scandalized, as a place, although the matter did not come under their immediate cognizance, at length began to take active measures in relation to it. All day Sunday there was a sort of half stifled hum heard through the village. The bells as usual called people to public worship, but not as usual was the solemnity of it regarded by the great mass of the people. Many, to be sure, went to meeting; but many did not appear to hear after they got there. Some thought ministers were such wicked creatures, they did not want to hear them; and some too just to condemn all, for the sins of one, endeavoured to listen with reverence, while their thoughts, in spite of themselves, would wander after him, who in their mind was guilty of this foul deed, and at this very time calling *sinners* to repentance.

Oh you! upon whom the authorities of the church, and the partiality of man, have conferred the envied distinction of speaking in your Master's cause, of being ambassadors for the greatest and highest of potentates, how great is your responsibility! a stain upon that spotless garment who shall wash away? If you are defiled by abominations, the destruction of your own souls is the least evil of which you are the cause.

All day, little knots of citizens were seen gathering at the corners of the streets, and even at the meeting-house doors, discussing the subject of the murder, though in an under tone of voice. Upon separating, they were invariably observed to shake their heads and walk away sorrowful. No active measures were, however, taken until morning; when a few citizens met in the street, and agreed upon having a meeting at the Lyceum Hall. A boy was sent about the streets with a bell, to notify the people to assemble, and very soon after the hall was filled to overflowing. Upon motion, a committee, consisting of five gentlemen, some of the most respectable persons in the place, was appointed; who were directed to "meet the Coroner and jury of inquest, who, it was understood, were that morning to be in session, and disinter the body for further examination; and if, upon examination, they should believe a murder had been committed, and upon having the evidence that some person was implicated in the murder, they should proceed to aid and assist the authorities of Rhode-Island in having the subject properly investigated, and in prosecuting it to a final issue."

At this meeting too, another and larger committee was appointed to collect and report to the first named committee, "any evidence or circumstance that might come to their knowledge, having a bearing upon the case." It was resolved that the truth should, if possible, be elicited in this search; and that they should report every thing of a favourable nature respecting the accused, as well as that which should appear unfavourable. Another meeting was subsequently held to make provisions for defraying the expenses of this committee.

It is said by the friends of Avery often, that he gave a manifest proof of his innocence in remaining in Bristol till the warrant came, and not fleeing or shewing any difference in his manners. The fact was, that he did not know anything was suspected of him, except his being the seducer of the girl. Mr. Bidwell, to whom Doct. Wilbour had, as before mentioned, related the conversations of the deceased, had proceeded immediately to Bristol and communicated with Avery, and had stated to him he was suspected of being the betrayer of the hapless girl. Avery and his friends got Mr. Bartlett, the stage driver, and a Methodist by profession, to go to Fall River and see how matters stood. In the mean time, Avery kept his house, walking it, as was said, in a state of very great agitation. He did no preaching that day. Bartlett proceeded to Fall River, and went in search of Doct. Wilbour, who was from home, visiting a patient. He followed him, and met him returning not far from his house, which they entered together. Upon going into the house, the Doctor perceived J. Durfee and another man from Tiverton waiting for him. Aiding Bartlett into the parlour, he went out to see them. They informed the Doctor that the warrant they had got was informal, and that it had been decided to apprehend Avery; and they requested him to go over the line and complain of him. This the Doctor refused to do, because he thought it was not his business; "but," observed he, "if he is not apprehended soon, he will be off. Here is Bartlett in the other room now, come to see how the business stands; and he will not get out of the place without finding out he is suspected of the murder." One of the gentlemen then proposed they should proceed immediately to Bristol, and have him put under arrest until the succeeding day, when a proper warrant could be procured; and begging the Doctor to keep Bartlett as long as possible, they departed, and in a few moments were on the way to Bristol. In the mean time, the Doctor apologized for delaying conversation until he had dined, after which he recounted the particulars of his conversation with the deceased to his interrogator, and concluded with the question, "and do you know that he is suspected of the murder too?" Amazed, the messenger answerd, "no:" upon which the Doctor assured him of the fact. Of course, he did not wait long after this, but hastened to convey the alarming intelligence to his employer. However, long before his arrival at Bristol, his friend and brother was under arrest.

It seems scarcely possible Avery could have refrained from preaching on that day merely from delicacy, because he had heard it was suspected he was the betrayer of the deceased girl, when he thus perseveres in it at the present day. However, Bartlett stated he was then very much disturbed and distressed in mind indeed, and that "he did not know when he had been kept from the house of God before."

Nothing was done hastily; the jury of inquest were very slow in their operations; and it was not until several days after the murder that Avery was arrested; and he probably might have escaped even that, had not new circumstances continually come up calculated to strengthen former suspicions. For instance, the other piece of the broken comb was found on the same back route to Fall River; fitted the first piece with which she was buried, and both were sworn to and identified as hers by the person who mended it and the people where she boarded, who, with the persons who worked next to her in the factory, deposed that she went out about six in the evening with it whole; changed her dress for one better; went in good spirits; and was exceedingly anxious to get leave to go out at the hour of six: had spoken of an appointment several days before to the daughter of the lady where she boarded; said she "did not care how many days it rained, if it was only fair on that day," 20th of December; shewed the pink and yellow letter which were afterwards found in her trunk to this young lady, who identified them; the white one also, with which she returned from the Post Office, on the 8th of December. The lady did not read the inside, but looked at the post marks and hand writing and was able to testify to them.

Those letters corroborated the statement made by Doct. Wilbour. The first of these letters, written on yellow paper, was dated, Nov. 13th 1832, and was as follows.

"I have just received your letter with no small surprise, and will say, I will do all you ask, only keep your secrets. I wish you to write me as soon as you get this, naming some time and place where I shall see you, and then look for answer before I come; and will say whether convenient or not, and will say the time. I will keep your letters till I see you, and wish you to keep mine, and have them with you there at the time. Write soon—say nothing to no one. Yours in haste."

They observed that he says, "I have just received yours;" and upon examining at the Post Office, Fall River, it was found there was one letter mailed for Bristol on the day preceding that addressed to S. M. Cornell, viz. on the 12th. But who it was for had escaped their recollection, if they observed at the time. Again, there was a letter on pink paper, addressed to the deceased, which a Mr. Orswell, the engineer of the King Philip, (a steamboat plying between Fall River and Providence) deposed was given him by Avery, in person, to deliver to Sarah Maria Cornell, near the last of November, while the four days meeting was holding among the Methodists at Providence. This letter too appeared to be in answer to one written not long before; and on the 19th of November the Post Master recollected that on that day, while making up the mail, he heard something drop into the letter box after he had cleared it; and upon looking, saw two letters, one for Bristol and one directed to Mr. Rawson, brother in law of the deceased, South Woodstock. This letter was afterwards produced by Mr. Rawson. His impression was, the other was directed to Avery; remembered distinctly it was for Bristol: and as it was ascertained he was correct about the first name, the committee could have no doubt about the other.

But so extremely cautious were they to go upon facts, that they delayed their proceedings until Orswell went up the river and saw Avery, to ascertain to a certainty, whether he would recognize the man who gave him the letter for that person. This letter, the one mentioned when speaking of her communications, to Doct. Wilbour, was as follows.

Providence, Nov. 1832.

DEAR SISTER—I received your letter in due season and should have answered it before now but i thought i would wait till this opportunity—as i told you i am willing to help you and do for you as circumstances are i should rather you would come to this place, Viz Bristol on the 18th of December, and stop at the Hotel and stay till six in the evening and then go directly up across the main street to the brick building near to the stone meeting house where i will meet you and talk with you—when you come to the Tavern either enquire for work or go out in the street on pretence of looking for some or something else and I may see you—say nothing about me or my family—should it storm on the 18th come on the 20th if you can-

not come and it will be more convenient to meet me at the Methodist meeting house in Somersett just over the ferry on either of the
above evenings I will meet you there at the same time or if you cannot do either i will come to Fall river on one of the above evenings
back of the same meeting house where I once saw you—at any hour
you say on either of the above evenings when there will be the least
passing i should think before the mill stops work—this i will leave
with you if i come will come if it does not storm very hard—if it
does the first i will come the second write me soon and tell me
which—when you write direct your letters to Betsy Hill and not as
you have done to me *remember this* your last letter I am afraid was
broken open.

ware your calash not your plain bonnet you can send your letter
by mail   Yours &c. B H
     S M C.—
let me still injoin the secret—keep the letters in your bosom or burn
them up.

The white letter found in her possession, marked one cent postage, was as follows.

                                              Fall River Dec 8th
I will be here on the 20th if pleasant at the place named at 6 o'clock
if not pleasant the next monday eve.—say nothing.

With respect to this last, final, and fatal letter, upon examination,
it was ascertained that Avery had been at Fall River on that very day;
had been heard asking for paper in a store kept by a member of the
Methodist meeting; and that that man went into the next store to get
a wafer for him: could not recollect whether he wrote in the store,
but remembered hearing him talk about writing to the editor of a
paper in the village, (whom, upon enquiry, he did not write to.)
From thence he went in the direction of the Post Office, and the
deputy post master recollected, a few moments before the stage

started for Bristol, in which he went, hearing a letter drop: and look-ing at the moment saw Avery just withdrawing his hand from the box. He then looked, and took out the one cent letter addressed to S. M. Cornell, when the wafer was wet. That wafer was recollected as the one supplied by the lady next door to the store where the paper was supposed to be procured—remembered from its peculiar colour.

The first letter, the yellow one, was post marked at Warren; and on that day it was ascertained the accused had been there.

The other letter was written by Sarah Maria herself, and directed to her minister, Mr. Bidwell. It expressed much compunction for her sins, confessed herself unworthy of a place in the meeting, and requested to be set aside as unworthy, &c.

With all these concurring circumstances before them, it is most evident the committee could not, in conscience, take any other course than the one they did take. Now previous to the arrest, when the suspicions of the murder were first excited at Fall River, his friends (Avery's) consoled themselves with the assurance that Avery would be able to prove where he was at the time of the murder; and it being a very cold blustering day until towards night, they had little doubt it would be found he was at his own house. What then was their consternation to find, upon enquiry, that he had actually crossed the ferry, at Bristol, on the afternoon of that very day, and after being absent on the island until a very late hour in the evening had gone back to the ferry-house requesting to be set over, which Mr. Gifford, the ferryman, declined doing on account of the late-ness of the hour and tediousness of the weather. There had been a rough wind for most part of the day, and generally in that place there is a considerable swell for some time after.

Still the friends of Mr. Avery kept up a good courage, for they felt morally certain that being in a methodist neighbourhood near so many friends and acquaintance he could easily be recognized, and would undoubtedly bring proof of where he was. But when after the examination at Bristol, it was found that he could not bring a single individual who even thought they saw him on the rout he described himself to have taken, many who had trust in him before fell off. He observed he had been on a walk of pleasure and observation, walking about the Island towards the coal mines, near the Union Meeting-house, &c. &c. past brother such a one and sister 'tother one, crossed a brook, went through a white gate, saw a "man with a gun, and a boy with some sheep," and finally wandered back to the ferry somewhere about ten o'clock, of a cold December night, without any supper or appearing to think of any? (though travelling minis-ters are not apt to forget such accommodations.) No man with a gun, or boy with sheep, could be heard of in that part of the country from any body but himself, and no one saw him, through all that route: nevertheless the justices appointed to examine him at Bristol, after what they declared to be a "patient, laborious, and impartial examination of the subject," discharged him. The county of Newport claimed him as their prisoner in the first place, and it was not a legal examination, because the offence alleged against him was perpetrated in that county. But his friends were determined to have his examination there, and they had it. By this illegal and ill judged proceeding the State was put to the expense of another examination, besides some much more heavy ones. The inhabitants of Fall River called another meeting and entered complaint to a magistrate in the county of Newport. A warrant was issued and a sheriff sent once more to take him.

CHAPTER IV

Upon arriving at Bristol, the sheriff found the prisoner had fled. Thirteen days had been spent in his examination, during which time he appeared so firm and unmoved for the most part that it was thought there was no danger of his decamping. He had fled however, and left his character to take care of itself. Those who believed him innocent, had thought he would court a trial in order to free himself from the odium attached to him, which unless wiped off they knew must forever destroy his usefulness as a minister of the gospel; but when they found he had decamped and left his friends and partizans to fight it out in the best manner they were able, they were confounded, but for the most part wise enough to keep still; and had he never been found, as most people believed he never would, it is probable the point would have been conceded. But he was gone: and Col. Harnden, the person who went in persuit of him, was almost at a loss to know what to do. There seemed no trace of him to be discovered. But although the person of the accused apeared to be beyond their reach, his character was not; and this flight, disgraceful and unmanly as it was, put the finishing seal to it.

Matters seemed so well arranged with respect to the reverend fugitive, that it would have puzzled wise heads to have known which way to look for him. But the indefatigable Col. Harnden was not to be daunted or disheartened in the cause he had undertaken. He had been one of the committee appointed to examine into this affair by the inhabitants of Fall River, and had satisfied himself that the accused ought at whatever cost to be brought to trial. He therefore commenced a most laborious and arduous search, and after traversing hundreds of miles backwards and forwards, in three States, having as he believed got on a track of him, he finally succeeded in his search, finding him in a remote neighbourhood in New-Hampshire, at the house of a Mr. Mayo. He was indebted at last to the sagacity of a baker's boy, who related a story of Mrs. Mayo being accused of some misdemeanor in the meeting, and Avery being sent for to plead her off, which he succeeded in doing, and saved her from the censures of the meeting—an evil of no ordinary character, if we may judge from the manner of handling the character of the deceased—and the lad thinking according to the old saying that one good turn deserved another, thought it must be he was concealed at that house. Upon arriving at the house, Mr. Mayo denied his being there, but observing his wife glide out of the room, Mr. Harnden followed her, and found Avery hid, pale and trembling behind the door of a chamber, evidently fitted up for his concealment, having the windows completely darkened, with lights and fire wood laid in, and all the comforts of life in abundance bore witness to the gratitude of her who held him in such gentle durance: pity that such comfortable quarters should have been disturbed by the intrusion of such unwelcome guests. Mr. Harnden returned with him through Boston, where, as in several other places, he, like other great characters, received the calls of his friends, the Methodists: Divines and all flocking to pay their respects—giving him the right hand of fellowship, &c. and having several "comfortable seasons of prayer,"* &c. with a man then laboring under the strongest presumption of being both an adulterer and murderer—of a man caught in the very act of hiding himself from the ministers of justice. "O tempora! O mores!"[1]

The authorities of the county of Newport, after examination, bound him for trial and he was indicted for murder by the grand jury, and the first Monday in May assigned for his trial. The interval

---

* Who can wonder that infidels should be strengthened by such things as these? what a farce does even christian worship appear when prostituted to secular purposes.

---

1. "Oh, what times! Oh, what morals!" Cicero, *In Catalinam* 1.1.2.

between the March term of the Supreme Judicial Court for the county of Newport, and the first Monday in May, was a busy one; scouts were out in all directions, and oh the racing and chasing there was to look up witnesses. Turnpike corporations and tavern keepers reaped a golden harvest during those two months. There was scarcely a factory village within a hundred and fifty miles but what underwent a thorough examination. The deceased it appeared, had been a moving planet, which she accounted for in one of her letters to her friends, by saying, "she belonged to a people who did not believe in staying long in a place." She seemed to have adopted for her motto, the text, that "here we have no continuing city;"[2] and she adhered to it in the spirit and the letter. Poor unfortunate being! she did not realize the danger of changing neighborhoods so often, nor know that it was safest for people to stay where they are best known, and where slanderers make out to live upon one old story for a thousand years, but transport it into a new neighborhood and ten thousand will immediately be added to it. She probably had never read that admirable fable of the Fox, who was advised to remove on account of the swarm of flies who beset him, and who wisely chose to remain where they might after a time get gorged with his blood, rather than to encounter a fresh and hungry set, when he should be robbed of every remaining drop of it.

In the mean time public indignation could not wait with patience for the issue of the trial, and from time to time it would speak out through the medium of the papers. This the methodists termed "persecution," whatever it was it is certain that much of it was provoked by their own imprudence in continually and loudly asserting his innocence, and the violence with which they endeavored to bear down public opinion, as well as their ridiculous fidgetting about the safety of his person, and his personal accommodation, through all the stages of his travels. Had Avery constituted solely in his person

---

2. Heb. 13:14.

the palladium of their rights, they could not have guarded him with more jealous care. They pretended to discover in the natural curiosity of the populace to see one who had become the lion of the day, a conspiracy to mob him; and at once took the responsibility of his flight upon their own shoulders, averring it to have originated in their fears for his person, and expressing terrible apprehension lest the Fall River folks should take justice into their own hands instead of waiting for the slow remedy of the law. The disgrace of flying from the persuit of justice, they affirmed belonged to them, having as they said persuaded him off and conveyed him to a secret place, against his own judgment. This last assertion may well be believed, viz. that "it was against his own judgment;" as that, if he had any, must have told him that his flight, under such circumstances, amounted to a strong presumption, if not to a confession of guilt. That he had fears cannot be doubted: he might have been in the situation of Trumbull's hero,

> Who found his fear of tar and ropes,
> By many a drachm outweigh his hopes.[3]

Their fears however of the vengeance of the "Fall River folks" were entirely without foundation, since nothing was intended but to bring the accused to a fair trial, and if his friends knew of such resources as they boasted of, they ought to have been the last to be afraid of that. But the important sixth of May arrived, and headed by an army of preachers, stout muscular men as ever took the field, followed by a company of women as a "corps de reserve"—and flanked by a hundred and sixty witnesses—the force of the prisoner

---

3.   John Trumbull, *M'fingal* (1782), canto 3, lines 435–36. In this highly popular mock-heroic poem of the American Revolution, Squire M'fingal, a Scottish-American Tory, loudly proclaiming his loyalty to the British crown, is assailed by a mob of rebels, tried, and sentenced to be tarred, feathered, and ridden through the streets—a common punishment of the day. Afterwards, the dispirited M'fingal realizes that war is inevitable. Williams slightly misquotes the actual lines: "He found his fears of whips and ropes / By many a drachm outweighed his hopes."

made its appearance. Newport swarmed with people of every denomination—curiosity was on tiptoe. There was a deep anxiety that truth should be brought to light by the friends of justice and humanity—and a restless and watchful one with others, to prevent if possible its developement.

The trial came on, and the prisoner was produced. He was a middle aged man, tall, and of very stout frame, and a face that might have passed for good looking, had not a certain iron look, a pair of very thick lips, and a most unpleasant stare of the eyes, have taken much from the agreeable; however it was agreed on all hands that notwithstanding these blemishes, he would almost any where pass for a tolerable good looking man, and moreover "looked like no fool;" or to use the language of the spectators, "looked as though he knew more than he told for." He was charged with three counts in the indictment. first—"for choaking and strangling the deceased." Secondly, "for tying her to a stake," and thirdly, inflicting various wounds and bruises on the deceased, calculated to cause death; or at least that must have been their meaning, though it was worded in the indictment, "of which she instantly died," but as no person could die twice, we presume this must have been the meaning. The prisoner of course plead "not guilty." The difficulties experienced in the formation of a jury were greater, it is believed, than were ever known before in any court in the United States, so strong was the presumption of the prisoner's guilt that it seemed almost impossible to find a man who had not made up his mind, and this mind was pretty rudely and unequivocally expressed by all on the spot: some few declared their feelings to be perfectly neutral, but one only solitary instance could be found of a man who said he had formed an opinion rather favourable to the prisoner; and it was not until after one hundred and eight were challenged that a jury could be found: the difficulty was materially increased by the prisoner's counsel, who in this as well as in every part of the trial seemed determined to carry every point by what is called management, and who fought the ground inch by inch—with so little apparent reverence to the authorities of the law that many a native of Rhode-Island blushed to hear the highest court in his state dictated to thus by a Boston lawyer.

As there are many, probably, who read this, who have never read the trial and never will, and some who will not even permit that document to come into their houses, we shall endeavour to give a summary of the evidence, though in a very brief and perhaps superficial manner; without going into the whole revolting particulars.

First then, the case was stated in a clear light, by D. J. Pierce, Esq. of Newport, the witnesses were then sworn. The fact of the death of S. M. Cornell was then proved, and of her appearance when found, as presumptive evidence she could not have hung herself, that she was taken down with the utmost care, rolled in a blanket and laid on straw in a horse wagon, and carried over a smooth road to the dwelling of Mr. Durfee, so that none of the bruises could have been inflicted after death. Here followed the testimony of the women who laid her out, and of the physician who examined her, the first and second time, for she was disinterred the second time on the 25th of January, when a more complete examination was had: this to be sure was nearly or quite a month after her interment, but it was in the coldest part of the year: she had been laid in a dry and marly soil, was frozen when she was buried, and the earth frozen that was thrown upon her, and the physician deposed that there was little alteration in her from the first examination. Every succeeding one brought to light new barbarities, and imagination sickens at the idea of the cruel butchery which this most unfortunate girl must have undergone, previous to her being strangled. No person could hear them unmoved: the very judges, though used to the delineation of crime, and pictures of violence, wept upon the bench; yea wept like children, at the description of her mangled person. We question whether the mere bodily sufferings of any one woman ever created such excitement, since the death of her whom the Levite cut in pieces and sent to all the coasts of Isreal,[4] which caused the death of more than forty thousand persons, and the extermination of a tribe.*

The circumstances of the letters were sworn to, and half a sheet of paper found in the store where the letter of the 8th of December was supposed to be written, which exactly matched the one of the letter, both the water mark and even the very fibres of the paper.

It was proved that the prisoner left his home on the 20th of December, without any good reason, without informing his family where he was going or assigning any excuse for absenting himself, that he had refused an invitation for that day to visit a Methodist lady, without giving any reason; that no person had been seen on the route he pretended to have taken on that afternoon; but that a man answering his description exactly was traced step by step all the way to Fall River, even to the very stack yard. One man, Mr. Cranston, at Howland's ferry bridge, swore to his identity. Mr. Lawton, the man on the Tiverton side, remembered a person of his exact description passing at the same hour, three o'clock. Mr. Durfee had been blowing rocks quite near the stack yard, and saw a man standing, and looking about with his back towards him. Abner Davis, at work there, saw the same man sitting on the wall, and upon his proceeding in the direction of the rock where they had just laid a train of powder (the direction of Fall River) called out to him, when he stopped. Both of their descriptions of clothes, person, &c. agreed with that of Avery, and upon seeing him they felt convinced he was the person, but as they did not see his face, could not swear to his identity. William Hamilton passing this spot about a quarter before nine in the evening, heard sounds as of stifled groans of some

---

* What mighty despotism, what scheme of bondage, what film of ignorance and fanaticism, what system of ecclesiastical tyranny, may not the death of this woman be intended to break?

---

4. See Judg. 19–20.

female in distress. The sounds appeared to proceed from the spot where the first piece of comb was found. He rose the hill and stopped, when, hearing nothing more, went on. One Ellinor Owen, who lived within sight of the place, about a quarter of a mile distant, testified to hearing screechings from that direction at half past seven in the evening. The cord was identified as belonging to some bags that lay in a cart of Mr. Durfee's within a few rods. A man answering his description went into the back room of Lawton's hotel, early in the evening on that day, and had a glass of brandy carried in. They did not know Avery, but upon seeing him, believed him to be the same person. Some person passed round the toll gate, at Howland's ferry, in returning, after it was closed, (after nine o'clock) by the beach. Their tracks were seen on the sand, where the water effaces any print once in twelve hours. The gate-keeper looked in the morning and ascertained some one had passed. He returned to Gifford's, at the ferry, late at night, about a quarter before ten, and said he had been on the island, on business; and to the question of the ferryman's daughter, if he "had a meeting that evening?" he returned for answer, he had not, but had "been on business to brother Cook's."

Nothing perhaps through the whole proceedings of the trial, examination, &c. gave more offence to the feelings of the public than the reckless disregard to character shewn by the prisoner and his friends in impeaching witnesses. This last mentioned witness, Miss Jane Gifford, was a young lady about eighteen years of age, of fair character it is believed as any other in the country: she had been a member of the Methodist class, and previous to this no one heard any thing to her discredit, but on this occasion they brought witnesses to swear "her character was not good for truth and veracity," though upon cross examination they were obliged to acknowledge "they had never heard any thing to her disadvantage previous to the Bristol examination." Several very respectable persons in the neighbourhood testified to her good character, among the rest Judge

Childs, who had "known the girl from infancy." They first spoke against her at the examination at Bristol, where she deposed to the fact. Another instance of this barbarity occurred at Bristol, whence it seemed an object to prove that S. M. Cornell was a wanderer and interloper at the camp meeting in Thompson: it was there mentioned she was seen with a Miss Rebecca Burk of Providence, and by her introduced to one or two methodists. Rev. Mr. Merrill was asked if she was not a leading member of their meeting? "No," he answered, "she was not a leading member there, that she had been set aside for *impudence*, and imprudence of conduct," or something like that, when the fact was that this same woman had been considered as a leading member in that society for more than twenty years, that she had been of great service in the cause of methodism in Providence particularly, and it is believed by many has done more towards building up the Methodist society in that town than any three persons who could be named, that she has given liberally of her substance towards the support of their meetings, though obliged to labor with her hands for her own support. The only thing they could have said, and that if fairly explained would have done her no harm, was some little disagreement between some of the members, of whom she was one, some time previous, wherein they were all what they call "put back," for six months, and at the end of that time restored, and every thing went on as before;* but there was nothing to affect her character. The words "impudence" and "imprudence" are generally understood to mean a great deal. But we are digressing.

Mr. Orswell, the engineer of the King Philip, gave a very clear and comprehensive evidence with respect to the delivery of the pink letter, by Avery, in Providence—"that he received it from the hands

---

* It is said to be their rule, where people dispute, to compel them to live in harmony. If so it is a good rule at any rate. People cannot always see alike, but they can refrain from disputing on their differences.

of Avery himself in person, who gave it to him between the hours of 8 and 9, or a little past 9, in the morning—that he received it with an express injunction to have it delivered as soon as the boat arrived, and gave him nine pence for carrying it—that he did not know Avery then, but went up to Bristol to see him, and recognised him at once, at his (Avery's) house, and to his anxious inquiries of 'what he meant to swear?' he replied, 'that to the best of his knowledge and belief he was the person.' Avery then put on his spectacles and asked him if he looked like him, and then turning to his friends asked them 'if he ever went out without spectacles?'" No notice appeared to be taken of this in court which was singular, as all the witnesses who saw him when he crossed the ferry (Mr. Pearce, Mr. Gifford, &c.) attest to his being without spectacles. It certainly amounted to proof positive of his artifice and dissimulation. He did in general, and for all that is known to the contrary, invariably wear his spectacles on going out, except this once, and on the fatal 20th of Dec. What could possibly be his motive for going on those two occasions without them, unless it was to disguise himself? Mr. Orswell did not positively swear to the day the letter was delivered him, but thought it was on Thursday. The letter was identified as the one he received, by the marks of his fingers which were smutty and oily at the time, and he recollected the manner.

With respect to the Camp Meeting, the source and origin, as she asserted, of her misfortunes, it was stated by a Mr. Paine, the young gentleman who carried her there at the request of her brother-in-law, "that he had seen her at various times during the summer, at the shop where she worked, while there, and that her conduct always appeared becoming and proper, and that he neither knew or suspected or heard of any impropriety in her."

The sister of the deceased stated that she (S. M. Cornell) returned from the Camp Meeting to her house, with a young man, an apprentice of theirs, Mr. Saunders; and that in September she confessed her fears of her situation to her, acknowledging her connexion with Avery at the Camp Meeting. The sister also swore to the fact that Sarah M. Cornell was free from any such embarrassment previous to that meeting. This was also sworn to by a Miss Lawton, a very respectable young woman in the family at the time, and who was her bed-fellow.

The brother-in-law of the deceased also testified to this confidence placed in himself and wife, and being troubled about it, he consulted his minister, Rev. Mr. Cornell,[5] and a lawyer, and that they both advised her removal to Rhode-Island. And further, both stated they never had the least reason to suppose she meditated self-destruction. That she had never, notwithstanding what had past, spoken reproachfully of Avery, but always mildly; and that her conduct at their house was perfectly proper.

Mr. Saunders, the young man mentioned, gave his testimony to the bringing of her from Camp Meeting—her behaviour perfectly proper, &c. in answer to the questions asked, and to his having put letters in the postoffice for her several times. "Were any of them directed to Bristol?" it was asked. "Yes." "Were the letters sent to Bristol before or after the Camp Meeting?" "Before." *Before* was the answer, and by what strange oversight this witness was not even interrogated we cannot tell; why after an answer that promised to them so much light on the subject, it was pressed no further is beyond conjecture. Many people previous to this had formed the conclusion that "Marmion and she were friends of old."[6] And that the betrayer had connived at her expulsion from the meeting, in

---

5. William Mason Cornell, a Congregationalist minister, was evidently not related to Sarah Maria Cornell.

6. Sir Walter Scott, *Marmion* (1808), canto 5, sec. 13, line 13. Scott's epic poem is a "romantic tale" set against the historic battle of Flodden Field (1513), where the English defeated the Scots under James IV. At the court of King James, Lord Marmion, the "hero–villain" of the poem, is entertained by the "syren," Lady Heron, the King's paramour. As she plays the harp and sings the ballad "Lochinvar," Lady Heron throws a bewitching glance to Marmion: "Familiar was the look, and told, / Marmion and she were friends of old."

order to conceal his own villany the better, and they thought they saw in this testimony of the correspondence with Bristol previous to the Camp Meeting a confirmation of their suspicions, that the interview with Avery at that meeting was concerted by letter: they therefore eagerly looked to see the witness further interrogated, but no such interrogation took place. *What he would have said, if interrogated*, belongs to another part of this story—and we hasten along with the trial.

The testimony of her sister and sister's husband was the only one which related directly to the interview of the deceased with the prisoner at the Camp Meeting. She had told them of an acknowledgment which she had given the meeting, about being unworthy, &c. &c.; that although she was then in good standing with them, it had not been returned; that Avery still had the paper in his hands and ought to restore it to her; that she had asked him on the Camp ground for that letter, and he requested an interview with her; that she met him in a retired part of the wood, where he asked her to be seated, which she complied with, and then asked him if he had got the letter with him? that he said "no", and then proceeded to take unwarrantable liberties, and that she made ineffectual resistance.

It seemed the friends of the prisoner had made great objections to the time it must have taken to walk the distance back from Fall River to Bristol ferry, on the night of the 20th; and two men now appeared and swore to the fact of travelling it recently in fifteen minutes less than it took the prisoner, according to all their statements: and further, it seemed the person whose route was traced all the way on that afternoon, was no slouch of a walker; for when he went over, a lady who saw him just on the Fall River side of Howland's ferry, remarked "that if that man kept on as he was going, he would get to Ohio before night." The evidence mentioned constituted the most important part, and pretty much all (though condensed into a much smaller compass) that was given. After the Gov-

ernment evidence was closed, the host of testimony on the part of
the prisoner was brought forward.

And first, there were six physicians, most of them of very considerable eminence in their profession; one or two holding professor-
ships in the medical department of some of our Universities. By
them they endeavoured to prove, first, that the deceased might have
hung herself; that there was a possibility of her hanging herself up
after she had strangled herself to death with the cord: next, that the
internal injuries discovered in her person on the second examination, viz. a little more than a month after the first interment, might
have been occasioned by decomposition. Those observed at the first
disinterment they did not attempt much to account for. Next, that
hanging was a very common death by suicide, and an uncommon
way of murder. (This was certainly a great discovery, and no doubt
edified the court and jury.) Thirdly, a long and most indecent
examination and discussion was entered into, to prove that the prisoner could not have been the father of the child which the deceased
was about to give birth to, and that her situation must have commenced previous to the Camp Meeting. This they ground on the
circumstance of the child itself, who perished with the mother,
being larger than common for that period of time. This seemed to
be the hinge on which they meant the case should turn; and for this
six physicians, some from a very considerable distance, were
brought together at a great expense, and a most lengthy and elaborate investigation was entered into, which for indelicate exposure
was probably never exceeded in any Court of Justice. It was a saying
afterwards that "the next age would have no need of physicians, as
every boy capable of reading would be perfectly instructed in all the
secrets of the Materia Medica—in the sciences of Anatomy and Sur-
gery, at least." However learned and elaborate it was, it is certain
that *one single question* put to those physicians, if properly answered,
as no doubt it would have been, would have put the whole to rest at
once, by overthrowing the whole theory they had been endeavour-

ing to establish. But the counsel for the Government happening to understand more of the laws of the land than the laws of nature, probably never thought of this test.*

Immediately after the testimony of the physicians, commenced the examination of a long string of witnesses respecting the character of the deceased, and here it has been shrewdly said the law was violated which provides that "persons shall not be compelled to give evidence against themselves." The whole sum and substance of the charges seemed to be taken from her own mouth, and women appeared on the stand and testified to things told them by the deceased of herself, "not fit for mortal ear to hear, or mortal tongue to utter." Such a repetition of village gossip—such a hunting up of old factory stories, and of legends long since, as one would have supposed, forgotten, (that is if they ever existed,) never was heard of before. Such a display of the amazing powers of memory too. That these witnesses were from different States, and therefore voluntary witnesses, relaters of scandals, of village gossip which never yet spared any one, and of which the good and the bad have sometimes to be equal partakers; betrayers of confidence reposed in them (by their own account) which if true proved at least that the deceased, however bad herself, considered them of the same stamp, for whoever heard of a loose woman pouring into the ears of a modest one the history of her intrigues? Whoever heard of such degrading herself by being in the confidence of a wanton? It appeared the obvious intent of such testimony to prove the deceased a perfect fiend, capable of plotting any atrocity and of carrying it through. But to what purpose it may be asked was all this directed? What possible bearing upon the case could such evidence have? It was not to prove the deceased good, but the prisoner bad, that the process was instituted.

---

* What sort of opinion those physicians had of the effect of their evidence may be gathered from the fact that one of them was heard to declare, not more than three weeks after the trial, "that he had no more doubt Avery killed her, than he had of his own existence."

It was not supposed that an immaculate, incorruptible being would have fallen a victim to the clumsy courtship and bungling attempts of a fellow who by the testimony of his own letters does not appear to have understood even the language he preached in, and a married man too. Why then this innumerable company of witnesses to blacken her character?

Why, as people generally understood it, it was for a threefold purpose. In the first place, the mere introduction of such a crowd of witnesses, the mere repetition of such a mass of evidence, was of itself sufficient to divert the attention, and confuse the intellects of any court and jury that ever sat. It had a certain tendency to throw dust in people's eyes, a phrase too well understood to need explanation here. And above all, its effect would be to turn indignation into another channel. This the wily counsel were fully aware of, and the doors once open to admit such evidence, they took care should not be speedily closed, but that every possible frailty or imprudence from the cradle to the grave should be hunted up and expatiated upon. It had the certain tendency to turn the public indignation from the murderer, whoever he might be, to the person murdered. And some were almost ready to exclaim, "No matter who killed her—such a person were better out of the world than in it—they have certainly done society a good service—whatever were the motives of the slayer, he has certainly conferred a public benefit." One person went so far as to say that "he did not think such a drab worth having a trial about!"

Persons of sense and discernment however there were who thought they discovered in this host of evidence great contradiction with itself. One of these evidences, a physician, related that she had come to him for advice, and told him she was a bad girl. Several witnesses too corroborate his testimony, and say that she told them of her calling on this doctor, and that he insulted her, and upon her repulsing him, threatened, unless she complied with his solicitations, that he would ruin her character with the meeting, and knowing therefore that it is esteemed a point of honour with physicians to keep all such things secret, and that he did immediately after the threat, as she said, and after the visit at any rate, say those things against her, they believed it done for revenge.* It was further proved that this physician made a demand of ten dollars which she refused to pay, saying she did not owe him more than half a dollar. Moreover one of these persons testified that the deceased told her she was doctored for a humour which originated in getting cold at a camp meeting—a thing by no means incredible to those who know the danger of sleeping in the night air, and on the damp earth.

The story of this insult she persisted in through all the subsequent trouble she met with, and they inquired if one part of her testimony was to be credited, why not all.

One witness testified to her going out of a factory with a string in her hand and she followed her, and really believed she was going to hang herself if she had not interrupted her.

Another witness of the Methodist Society testified that the deceased once told her that she attempted her life, and had not courage to go through with it.

Two persons, a tavern keeper and his wife, by the name of Parker, gave a most singular testimony. They averred that eight years before, a girl *calling herself* Maria Cornell, came to their house in the evening, being evidently in a situation no young woman would want to travel alone in, and "appeared much engaged in *the work of God.*" That was the expression. When two young men entered, and she immediately charged one with being her betrayer, and frightened him out of a sum of money to settle with her, and gave him a receipt; and that they all staid all night; and she came down stairs next morning looking entirely different; and the young man thus swindled took no notice of her altered looks; and they all went away, it would

---

* Others however swore she confessed this charge.

seem perfectly satisfied. This evidence was judged of great importance it appeared by the prisoner's counsel, by the manner in which
it was handled by them. By others it was received as exhibiting
inconsistences not to be reconciled. 1. That any young man would
put up with such an imposition. 2. That any one engaged in such a
fraud would so soon throw off the mask. 3. That persons so very religious as they evidently wanted to be thought, would tolerate such
transactions in their houses, and lodge the whole company after it,
the woman, whoever she was, and her paramour.

Four women and two men (Methodists) were then examined;
the women gave a history of such disgusting intrigues, as could
scarce be paralleled, which they said the deceased acknowledged to
them in the way of confession. That she appeared very penitent for
them, and one said "wept upon her neck until she was quite dis-
gusted with her." Two testified she had threatened vengeance
upon Mr. Avery for signing her expulsion from the meeting, and
that she said "she would be revenged on him if it cost her her life,"
although it did not appear she had any known cause of hostility
against him. No one could attach any blame to him, who being her
minister was obliged to act as the rules of the meeting required.
One of these fair swearers was one of those who went off with him
at the time of his flight from Bristol.

Another instance of that recklessness displayed by the prisoner's
friends of the character and peace of individuals, and perhaps the
most barbarous, was the trying to disgrace the character of her sister's husband, a young man of most unexceptionable character,
always known for his modesty, sobriety and piety, and who had not
seen the deceased for several years previous to the fatal summer—
the last of her earthly pilgrimage, when she came to visit them and
her aged mother, who resided in the family. They brought witnesses to say, that S. M. Cornell had told that her sister's husband
had loved her better than his wife, and that they had been as free as
man and wife; and one of the witnesses, a young girl, recited a long

piece of poetry which she recollected she said from reading it once or twice, and which Maria had said her brother addressed to her.*

The next company of witnesses were called to cover the time of the prisoner's being at the camp meeting, which if we recollect was three days, in such a manner as to occupy every moment of it, and make it an impossibility of his having any assignation with the deceased. In fact, if all this testimony could be relied on, the prisoner had not only no time for an interview with any woman out of their sight, but no time neither for the ordinary occasions of life, no hour for private devotion or any thing of that sort; wherever he went, it appeared from the testimony, there was some one at his elbow; if he walked or rode, or sat or slept, or eat or drank, or preached, somebody appeared to testify to the hour, whose memory was fresh with every particular. As one dropped him another took him up; if one left him another joined him at the same moment, until they fairly guarded him out of the premises, and out of the country. Had E. K. Avery been a State prisoner, suspected of treason, under one of the most arbitrary governments in the world, he could not have been more strictly guarded, and closely watched than he must have been, even if the whole College of Jesuits had been on the alert—besides having such fine memories that they could all remember so exact about every moment of time, and even the slightest circumstance respecting this man. Many argued, who heard this testimony, that this was suspicious; that it was a thing contrary to general experience—that among such a multitude, one person of no very extraordinary character for any thing, should be singled out as an object of remark, a point of observation,

---

* It did not seem sufficient that her almost distracted sister had to be dragged to Court, to hear this load of infamy laid upon the departed, but her domestic peace must be assailed, by suspicions endeavored to be infused of the fidelity of her husband, of the father of her children, now her only earthly support and consolation. "Oh," said she, when speaking of this transaction afterwards to a friend, "had I been at all addicted to jealousy, or had the least cause to be so, or possessed as weak a mind as they imputed to my sister, what might not the consequences have been. They might have broken up my family and perhaps driven me to distraction or suicide, but to disturb my peace in that way is beyond their power." Still we must suppose she was a very great sufferer in hearing such abuse.

a centre of attraction to which all eyes were turned, and argued from this very testimony, as well as the similar one of the four days meeting in Providence, that it was overdone, and would undoubtedly have a tendency to convince the court of the delinquency of the prisoner—the result however disappointed their calculations.

In the same manner they endeavored to cover the time of day Orswell supposed was the one that the letter was handed him. The Attorney General stated in his remarks, that so earnest had they been to cover the time, when the letter could have been delivered, that "they made out fifteen minutes more than there really was of it." It appeared from the testimony of the witnesses, that he was constantly with some of them except when he went from breakfast to the clergyman's, when the walk was accomplished in as short a space of time, as ever man walked it, and immediately appeared in another place, when the Rev. somebody else took him to brother somebody's, and instantly he appeared again in the methodist meeting-house, at the beginning of the meeting, precisely at nine o'clock; this they remarked by one particular circumstance, it is, that he did not open the meeting which he had previously agreed to do; for this omission no reason appeared from him or his friends, so that people were left to conclude, either that there was a mistake in the day, which Orswell did not swear to, or that he had slipped away a few moments before the meeting (eluding the vigilance of his sentinels,) and was too much fatigued to open the meeting after such a race; or that he excused himself from making the first prayer, in order to slip away while the people were on their knees and would not observe him; and as to other profane spectators, they would not have observed the circumstance of a man gliding in and out, where there is such constant ingress and egress.*

---

* This dodging about in Methodist meeting is believed to be nothing uncommon. The writer of these pages has a very distinct recollection of J. N. Maffitt, who used frequently while another minister was praying, to climb up and look over the house, to see who stood affected, and either go to such after, or have them brought up to *have the benefit of his prayers:* as it was not noticed as a breach of decorum, we conclude it is not uncommon.

While the evidence was taking, witnesses arrived post haste from Providence, to swear that they had just measured the distance from the methodist meeting-house to the steamboat wharf, and found the distance so great that it was impossible he could have travelled it that morning before meeting. So Mr. Orswell was completely sworn down. Nevertheless a little time after that trial was decided, a respectable farmer came forward and testified to seeing Avery when he delivered the very letter to Orswell. He did not know Avery at the time, but when the trial came to be published, accompanied with a striking likeness of the Rev. accused, this man, Mr. Angell, immediately recognized the person.*

Witnesses from the Camp ground were produced against S. M. Cornell, the deceased: one of whom testified she saw her slap a young man on the shoulder; another thought something might have been the matter with her, as she thought she walked different from other folks; another imagined something against her character because her frock did not quite meet together behind; another testified that it was said there were persons of bad character there, who were directed to be ordered off the ground. But although it was known the deceased was there, it appeared she was not molested. As to the character of Avery, a number of their witnesses were examined, all of whom testified to the faultlessness of his character; never heard but what his disposition was good; his character for every thing, good. Two of these witnesses, Methodist ministers by the name of Merrill, upon being cross examined, confessed he had been prosecuted for defamation in Massachusetts, but stated it resulted in

---

* The measuring the ground and deciding he could not have gone on account of the distance, reminds us of the trial of John N. Maffitt, whom a clergyman of unimpeachable character saw kiss his hand during service time to a lady in the gallery. The methodist conference went and measured the distance from where Maffitt stood to the gallery, and very gravely decided that the distance was so great that the witness could not possibly have heard the *report of the kiss!!!* and that their worthy brother must be innocent. [Maffit, a Methodist preacher in Boston, was acquitted of misconduct in 1823 in an ecclesiastical trial.—EDITOR.]

nothing to impeach his character, and that *the Ecclesiastical Council acquitted him of all blame.*

And what, asks the reader who has never read the trial and is unacquainted with the events of this story (unless Ecclesiastical Councils should take the place of Courts of Justice, and become the law of the land, and such books be condemned to be burnt by the common hangman and their authors to some modern "Inquisition,") what did they, the witnesses, say respecting the absence of Avery from Bristol on the day of the murder? and how did they manage to clear up the circumstance of the letters? of his being so unfortunate as to be in the very places, on the very day when the letters were dated, and of having letters charged him at the Post Office on the very day when the letters of the deceased must have reached him? Surely here must have been their strongest stand; and these things satisfactorily accounted for would not only have saved his life, but what is of more value, or ought to be to a christian minister, his character. Doubtless this must have been the ransacking for witnesses at the time the turnpike gates saw such hard service. This must have been the dodging in and out of every tavern, factory village and factory boarding house in the country.

No such thing, no such witnesses were brought forward, nothing of the kind attempted. Relying upon the protection of the law, that the accuser shall prove where the accused is, not he prove where he is not, the prisoner took possession of the strong hold, and saved the ship from sinking by throwing character overboard.

But surely, says the reader, they must have made a lame piece of work of it, if that were the case. For what purpose this array of witnesses to prove the deceased bad? that was what the Government wanted to prove: for good she could not be and be his mistress— her minister! a married man too! Why it argues a great degree of depravity, or infatuation, or destitution of reason. How did all these inconsistences of character in the deceased help him? It only made the probability of the case more apparent. Granted—but nevertheless this testimony, strange and inconsistent and contradictory as it was, was their fort, and upon this they grounded their defence of the prisoner. The counsel for the prisoner had sketched out a romance, not to be equalled by any thing we know or read of Spanish or Italian vengeance, and dressing it up in a most ingenious manner, presented it to the attention of the Jury. His argument was, that this girl, the deceased, was utterly bad, capable of any sort of wickedness; that she owed the prisoner a grudge for his share in turning her out of meeting, and that she had wreaked her vengeance upon him in this manner: first, by writing the letters or procuring them to be written and sent to her, and then by pretending he was her betrayer; and finally hanging herself after writing a billet, "if she was missing to inquire of the Rev. E. K. Avery."—that she had said, "she would be revenged upon him if it cost her her life," and accordingly had contrived this method and carried it into execution, and that all the rest was the effect of the heightened imagination of the Fall River folks; and the excitement he politely styled the "Fall River fever": and whenever in the course of his brief review of the evidence, he chanced to come across something remarkably tough, why, with a flourish known only to the people called lawyers, he would give it a toss, and get rid of it at once without any trouble, as easily as one would toss a biscuit into the sea. Never was the old proverb verified better than in this case, viz. "one bold assertion is better than a host of argument," and "two negatives is as good as one affirmative;" and we had like to have added the third, "a lie *well stuck to* is as good as the truth," but we leave that out. He attempted to establish it as a fact that the deceased was insane too, and yet that all this method was adopted in her madness: that she was capable of a plot of revenge deeper and of a more diabolical character than any ever related before of woman—a plot which, in conception and execution, surpassed all human credibility.

He was replied to by the Attorney General, Albert C. Greene, Esq. whose health at the time was not good, and whose arduous labours had during the trial much exhausted him; a gentleman of good law knowledge, of amiable manners, and feeling heart, but whose plain good sense was no match for the subtlety of his antagonist. His speech contained much sound reasoning; nevertheless, after a short charge from the chief justice, the jury retired, and on the next morning, at 9 o'clock, brought in a verdict of *not guilty,* having consumed four weeks in the trial.

Various opinions respecting the verdict of the jury prevailed, yet all felt it their duty to acquiesce in the decision of a legal tribunal, and no one had the least idea of molesting Mr. Avery after his discharge by the court. The Fall River people, who had behaved throughout most magnanimously, notwithstanding the hue and cry of the friends of Avery, that they were thirsting for his blood, and a deal more of that sort, were as content to let him live as any others. They however looked forward with certain confidence to his being deposed as a preacher. They could conceive of very great efforts to save him from the gallows, from the mistaken notion that the penalty was the disgrace of crime, and that his death would be thought to bring an uneffaceable stain upon the methodist order. When therefore his own people sat upon his case, as it was known they did not measure their decision by the fiat of the law, and that he did not, nor could not, satisfactorily account for himself, or clear up the affair of the letters, &c. it was believed he would be expelled from their order, or at least forever debarred from preaching—that if it were for their own character alone, they would not suffer such an outrage upon the feelings and common sense of the community. But to their amazement and that of others the "Ecclesiastical Council," as they style themselves, the highest tribunal among them at any rate, pronounced him perfectly innocent, and freed from all suspicion, and continued him in the service of his office. This outrage upon the feelings of society it is believed will eventually injure them more in the estimation of mankind, than it would have done to have had twenty preachers hung.

To leave digression and pursue the thread of the narrative—E. K. Avery was almost instantaneously hurried out of Newport, after the rendering of the verdict, and conveyed to his family in Bristol, and continued in his office, and weekly to hold forth to the people, followed by crowds whom curiosity attracted to hear him, so much more will that impel people than devotion.

The murdered, mangled remains of Sarah Maria Cornell still repose at Fall River, at rest we hope, from all further molestation. The generous and feeling inhabitants of the village wished to have placed a handsome marble monument over her remains, detailing the sad tragedy of her death, but this her relations objected to, from the fear that it would not be permitted to remain, and that the same interest which had been exerted to blacken her character, might be to destroy all records of the transaction. Her brother and sister Rawson therefore placed a small but neat stone at the head and foot of the grave, simply inscribed with her name and age. That lowly grave has been the pilgrimage of thousands from all the different sections of the country. It is in vain that the friends of Avery endeavour to place that unfortunate being beneath even the pity of the virtuous. Her own sex feel she was a woman, and as such entitled to their sympathies, the other, more generally inclined to compassionate female frailty, pity her with undissembled sorrow. Few have visited that spot without tears. There seems to be a spell breathing around that none can withstand: the effect is absolutely irresistable. It is a humble grave, in a solitary spot. It is the grave of a poor factory girl, but from that grave a voice seems to issue, noiseless as that still small one,[7] that speaks to the conscience of the sinner, but whose tones nevertheless sink deep into the heart. The author of these pages visited that spot, as well as the one where she met her fate, at a most interesting moment. It was on the evening of the first of July. The moon was then at its full, yet a kind of shadowy darkness hung over

---

7. For the "still small voice" see 1 Kings 19:12.

the spot, blending the outlines of the surrounding landscape so as to render them nearly indistinct. For some time I stood wondering, without dreaming of the cause, but upon looking up, discovered the moon was in an eclipse. There was a singular coincidence in it certainly, and it forcibly reminded me of the dark and mysterious fate of her who reposed beneath. I watched it as the shadow slid from the moon's disk, and I felt that confidence which I have ever felt since, that the mystery of darkness which envelopes the story and hides the sad fate of that unfortunate victim will one day be dispersed. The following lines were penned at the time and afterwards published in the Fall River Monitor. They are inserted here by request.

> And here thou makest thy lonely bed,
>   Thou poor forlorn and injured one;
> Here rests thy aching head—
>   Marked by a nameless stone.*
>
> Poor victim of man's lawless passion,
>   Though e'er so tenderly carest—
> Better to trust the raging ocean,
>   Than lean upon his stormy breast.
>
> And thou though frail, wert fair and mild;
>   Some gentle virtues warmed thy breast.
> Poor outcast being! sorrow's child!
>   Reproach can't break thy rest.
>
> On thy poor wearied breast the turf
>   Lies quite as soft as on the rich:
> What now to thee the scorn and mirth,
>   Of sanctimonious hypocrites.
>
> That mangled form now finds repose,
>   And who shall say thy soul does not,
> Since he who from the grave arose
>   Brought immortality to light.
>
> Poor fated one the day is coming
>   When sin and sorrow pass away—

---

* The stones with her name were not then up.

I see the light already gleaming
>   Which ushers in an endless day.

Where shall the murderer be found?
>   He calls upon the rocks in vain—
The force of guilt will then confound,
>   Alas the Judge! no longer man.

He calls upon the rocks in vain—
>   The adamantine rocks recoil,
Earth can no longer hide the slain,
>   And death yields up his spoil.

Where shall the murderer appear?
>   My God thy judgments are most deep:
No verdict can the monster clear
>   Who dies a hypocrite must wake to weep.[8]

---

8. The poem was written by Williams.

CHAPTER V

LIFE OF SARAH MARIA CORNELL

With the greatest care and impartiality the author of the following pages has collected together all the facts susceptible of proof relating to the life of Sarah Maria Cornell. Some of these were gained from her own family—others from strangers.

S. M. CORNELL was born in May 1802, in Rupert, Vermont. Her mother, the daughter of Christopher Leffingwell, Esq. of Norwich,* was a well educated and good principled woman, a daughter to one of the first families in the State. She had been carefully brought up and accustomed only to the best society. Unhappily, she contracted early in life an unfortunate attachment. Mr. Cornell was a person employed in one of the manufactories belonging to her father. Good looking and of pleasing address, he succeeded in captivating the affections of a daughter of his employer. Mr. Leffingwell was at first

---

* This Christopher Leffingwell was the direct descendant of that Thomas Leffingwell of Saybrook, Connecticut, who had the honor of rescuing by his bravery the celebrated Uncas, with his remnant of Mohicans, from the power of the Narragansetts, in the bloody war between the Indians of this last tribe and the new settlers, the English, about the year 1660; and who received afterwards, as a testimony of gratitude from that renowned warrior, the grant of land, by deed, of all that tract upon which the town of Norwich now stands. New-England is under lasting obligations to the name of Leffingwell. The circumstances were these.

Uncas, who with his band was fighting in defence of the whites, got hemmed in, in a place of imminent danger, at some distance from Saybrook, but found means to send a messenger to that place to ask the English there to come to his relief. Their whole force had left the place, in another direction, except those left to guard the fort. But Thomas Leffingwell formed the bold plan of conveying the whole band across into the fort, in the course of the night, in his canoe, and actually accomplished it; and when the ferocious Narragansetts came upon their post, in the morning, behold they were gone! all safely stowed into the English fort at Saybrook. This manœuvre turned the tide of war.

SARAH MARIA CORNELL
"Her countenance bespoke the possession of beauty in happier days—
but it was now clouded with care and shaded with grief."

very wroth, and made considerable opposition to the match, but upon being assured by his daughter that she was firmly and immoveably attached to Cornell and could never be happy with any other man, the old gentleman gave up the contest, and suffered the union to take place without further opposition. His daughter removed after marriage to Vermont, where her children were born; and here she was destined to taste the bitterness of an ill assorted union. Her husband it seemed had formed the design, and it very soon developed, to be supported from his father-in-law's funds, which were supposed inexhaustible, and himself to be a gentleman at large. In pursuit of this determination he worked upon the feelings of his wife to get her to draw money from her father. Mrs. Cornell, who was one of those gentle, unresisting characters that knew not how to contend, suffered herself for some time, though sorely against her feelings, to be influenced to this, and repeatedly drew large sums of money from her indulgent father, to supply her husband's demands, until at length the old gentleman resolutely refused to advance any more; upon which Cornell carried his wife and children to her father's house, and leaving them, quit the country, and relieved himself forever from the task of supporting a woman whom he had probably married without the least sentiment of affection whatever, and abandoning the children in their helpless infancy, whom the laws of God, and the laws of the land both required him to support. What was the situation of Mr. Leffingwell's estate at his decease, we do not know, or whether he supposed he had bestowed enough upon this daughter; but certain it is that although the rest of the family were in easy circumstances, if not affluent, she and her family were poor, and she and her children found a home with some of their relatives, and appear to have looked chiefly to their own exertions for support. They were separated, being all brought up at different places, and not even knowing one another for several years. The unfortunate girl who is the subject of this memoir was in the same house with her mother until about eleven years of age. She then went to live with a Mrs. Lathrop of Norwich, her mother's sister. With her she continued until fifteen years of age, and then went to learn the tailor's trade, where she staid two years, and then for a time resided with her mother in Bozrah, a short distance from Norwich, working at her trade.

During her residence at the house where she learned her trade, her mind appeared for the first time called up to attend to religion. There was at the time a great reformation, as it is termed, in the neighbourhood—that is, there was a great stir about religion, and much going to meeting, and many professing, of which number doubtless many continued steadfast; but in a time of such general excitement it is known there is a great deal of self-deception. The quick feelings and sanguine temperament of S. M. Cornell were calculated to mislead her, and it was not long before she rushed with the multitude to the altar of baptism, joining herself in christian communion to the congregation of the Rev. Mr. Austin, a Calvinistic Congregationalist. No reproach can with justice attach itself to a clergyman in such cases, unless they are hurried into such a profession without any time for trial, which was not the case in this instance. Man cannot look into futurity and tell who will prove steadfast and who will not, and if a rational person makes a good profession of faith, and avows a resolution to lead a christian life, the minister is bound to receive them, unless he knows something in their present character and conduct at variance with their professions. For two years she continued steadfast, and it was said a bright example in outward conduct; yet nevertheless the seed had fallen on stony ground, where the earth was not of sufficient depth to foster it. A season of declension succeeded it. Lightness and vanity again took possession of her imagination. A passion for dress at this time seemed to be a predominant feeling, and that passion she was obliged to set bounds to, because she had not the means of gratifying it.

It was at this unfortunate season, the only one it is believed in her existence when the same temptation would have had the same

weight, that her mother brought her to Providence. Her older sister
lived there with a relation who had brought her up; and these two
sisters, separated for many years, had long desired a reunion. That
wish, so natural, was at last indulged, and like most of our earnest
desires for earthly gratification, indulged to their mutual sorrow.
Introduced for the first time since childhood into the temptations
and allurements of a commercial town, those feelings of childish
vanity, and love of dress, and show, and ornament, which had been
growing upon her for some time, seemed completely to get the mas-
tery—and being often in the shops where those articles for which
she had so long sighed presented themselves before her—she at
length possessed herself of some of them, trifling indeed in amount,
but destined to prove her entire destruction in this world as
respected character and every thing else. Though the whole of these
articles purloined in a moment of lightness, of thoughtlessness and
temptation, did not exceed in amount but a very few dollars, it was
immediately discovered, and the avenger was close upon her heels.
Unused to crime, her manner at the time was so singular and agi-
tated as to excite suspicion in the store, and she was followed to the
house of one of her relatives, where the articles were found—not
exceeding five dollars in amount, and several very small trifles
beside, which she immediately told of and where she got them, and
her friends sent them to the gentlemen, and offered to pay all dam-
ages, &c. to both; they exacted nothing however but the amount of
the goods. The grief and agitation of the poor girl vented itself in
repeated fits of hysterical laughing and crying at the time, and in the
bitterest self-accusation afterwards, when she seemed fully to realize
what she had done, and could those gentleman have known the
effect that disgrace was to have upon her future destiny, doubtless
they would have preferred to have lost ten times the amount rather
than have exposed her. Be that as it may however, the fact that she
did purloin these articles is certain, and I have it in express charge
from her nearest kindred, her kind brother and sister, not to attempt

to conceal it, but in every thing as far as I can discover the truth to make it manifest. They knew of this delinquency in their sister by her own confession; she did not attempt to deceive them, and they knew of no other instance of the kind of her offending; they know by the same means, viz. her own confessions, of her intercourse with Avery, and they know of no other person with whom they believe her to have been criminal. But to go back to the story.*

The open, candid manner in which they had behaved, themselves, and the keen distress of the offender herself, certainly induced them to hope she would not be publicly exposed, but by some means or other it was immediately communicated to town and country. For this they were not prepared, far less did they anticipate that this circumstance would be brought up in a court of justice, eleven years after, to prove that she killed herself, to be avenged on a man who had exposed her misconduct, when she had not even shewn resentment towards them.

That this was the only sin of the kind—the only instance of dishonesty that could be brought up against S. M. Cornell, must be believed by every one who ever saw the famous trial of S. M. Cornell, denominated on the title page, "Trial of E. K. Avery." For had there been another thing of the kind known against her—a wit observed as "heaven, earth and hell were ransacked for witnesses," it must have made its appearance. On the contrary, she was afterwards often remarked for the punctuality and exact regularity of her dealings. The writer of these pages knew a milliner with whom she had very considerable dealings at Lowell, and to whom she was often indebted, and who remarked "that she was the most punctual per-

---

* Some of the people so violent in denouncing this poor girl, at the time, were running crazy after a new preacher then in town, who, they affirmed, was one of the greatest saints living; as he had done every thing bad—murder excepted. Among other things, he had been a thief, they said. Not thinking that any particular recommendation in a preacher, we had not the honor of hearing him; but we recollect remembering, at the time, the old adage, "one man may steal a horse, while another man cannot look over his shoulder."

son in the payment of her debts she had ever known, as she seemed to have a principle of honesty about discharging a debt the very day she had promised the money, and always bore in mind the exact sum she owed."

It appears that the connexions of S. M. Cornell generally, with the exception of her mother, and her kind hearted sister, meant to make her feel the full extent of the offence she had committed. It does not in the general way require much to set rich relations against poor ones—but here was ample room for feelings of superiority over *poor*, fallen human nature. Some of her connexions shut the door in her face when she called to see them afterwards—and for the most part they manifested a very *proper detestation* of her offence, by displaying *proper resentment*. She returned to the country and resumed her employment, but the story got there before her. She had relinquished her former employment of tailoring and gone to work in a factory. Here she was now regarded with a degree of suspicion, painful in the extreme to a person of her natural pride, and she quit the place and went to another, but being dissatisfied with the employment, again resumed her sewing, and went to live with a merchant tailor in a neighboring town; she continued in her employment some months, when the story reached the family that she *had been talked about*, which caused them to watch her with scrupulous regard. There was a young gentleman then in the neighborhood who used to go often into the shop, and frequently sit down by her and converse, sometimes in an under tone, and sometimes he would invite her to take a walk of a pleasant evening, and she would go with him. This circumstance, as he was a young and unengaged man, and she very pretty, would probably of itself have caused no suspicion, had not the saying that she had been talked about been so often repeated. She did not board with the family who employed her, but in the family of a respectable physician on the other side of the way: and being convinced by the circumstance just related, joined to the saying that she had been "talked about," though they did not exactly know for what, that her character was not good, the wife of her employer took it upon her to dismiss her; and sending for her to come in, begun by accusing her of "imposing herself upon their society when her character was not good;" and having said all she judged necessary on that head, she formally dismissed her from their employment. During all this time the poor, persecuted girl only opposed tears to the reproaches heaped upon her. She knew that she had, by one indiscretion, by one violation of that command, "thou shalt not covet any thing that is thy neighbor's," brought reproach upon her good name; and she probably thought they knew of it, and said nothing because she could not bear to hear it named. She only asked permission to remain until next day, when the stage would pass, which was granted. "To this day," said the lady who had vented these reproaches, "to this day, my conscience reproaches me for the harshness with which I spoke to her, when memory recalls the tears she shed, and her meek, forbearing manners, and I must say, that she had the meekest temper, and one of the mildest and sweetest dispositions I ever met with." She added, that that very night a relation of theirs who was then very ill in their house, was distressed for a watcher, they having sent half over the neighborhood for one without success; which S. M. Cornell hearing of, immediately offered to watch with her, and though they were ashamed to accept of her services, they were constrained to; and that she was so kind and attentive to the sick, that the woman after her recovery often enquired after her, saying, "she was the kindest and best person to the sick, she ever saw."

From this place it appears she went to Slatersville, Rhode-Island, and commenced working again in the factory; soon after which, a Mr. Taylor, a Methodist, commenced preaching there, and here again there was a great stir about religion. Mr. Taylor was one of their popular preachers—there was a great reformation, and S. M. Cornell, who had for some time given up the idea that she had ever possessed religion, was once more awakened; and having, by some means or other, become persuaded that immersion was the only

Scripture way of baptism, felt desirous to be rebaptized. After a profession of faith and going through all the preliminaries, she was accordingly immersed; and the Methodist meeting, who profess to believe that water administered in any form, in the name of the Trinity, is baptism, and who baptize in both ways themselves, had no hesitation in rebaptizing her. However, that is of minor consequence to what followed. She continued in fellowship with them, it appears by her letters, during her stay in Slatersville, which must have been over two years; for she staid there until the factory burnt down, and then of course had to depart in search of employment. With several others she removed to the Branch factory, a few miles off. Here she staid until the water becoming very low, there was not steady employment, when she removed to Millville, to the satinett factory. From this place, only about a mile and a half from Slatersville, it will be seen by her letters, she attended her beloved Methodist meeting at Slatersville, and appears to have felt great joy at finding herself so near there again. No person can read her letters and suppose she feigned what she wrote. Just before her leaving Smithfield, i.e. Slatersville, Mr. Rawson, her brother-in-law, went and carried her brother, who had been absent several years at New-Orleans, to visit her, and inquired of the family where she boarded, "how Maria got along?" "Very well indeed," was the reply, "and much engaged in religion," they added, "and set a very good example."

While at this place her zeal in the cause of meetings continued. It appears she was in the habit of walking down to Slatersville, on all occasions, to meetings; and that in the prayer meetings as well as those for exhortation, she usually took a part, and was called an active member. We do not know whether she was censured at this time, but this fact we do know from letters in our possession, that she was in the habit of corresponding with methodist sisters at this time, and subsequent to it, who were highly spoken of for piety and consistence. We have some directed to this last place, and they address her as "worthy sister," and solicit an interest in her prayers.

It had been the intention of Maria (by that name she was generally called) to return to Slatersville as soon as the new manufactory should be completed, and never to leave the people with whom she was connected there until death, but unfortunately the works did not keep pace with her impatience; she disliked the woollen factory where she worked at Millville, and one of the girls who had been a favorite companion and sister in the church persuaded her to go to Lowell, and declaring her determination to go there first, which she did not however do immediately, as Maria came to Providence to visit her friends, or more particularly to visit her dear mother; and after staying some little time in Providence and Pawtucket, received a line from her friend urging her again to go to Lowell, and naming a place on the road, where they would meet on a certain day provided she would comply. The place was in Dedham, and here they concluded to remain, but there being no methodist meeting, she became discontented, and after four weeks residence there proceeded to Dorchester. What caused all this delay in going to Lowell is not known, unless some guardian spirit interposed and delayed her progress to the place which was to consummate her destruction. During the time of her sojourn in the towns already mentioned, at several different times she received attentions from some young man, who she thought and others thought wished to marry her. Many young men make a practice it is well known of amusing themselves at the expense of young women who are apparently without friends and natural protectors to call them to account for such baseness and compel them to act honorably. S. M. Cornell had the curse of beauty, and she was not without admirers. She was naturally of an affectionate and confiding disposition. Her manners too, all partook of that character of fondness for which she has been so unjustly censured. She loved her mother and sisters, and her letters bespeak any thing but a depraved heart. It is an indisputable fact that an abandoned woman is without natural affection, and we see that she was the very reverse of this. Her letters she did not even know would be preserved. Little could the poor, unfortunate girl have dreamed of the use here made of them: they were only to meet the eye of her sister and her aged and bowed down mother. It seemed as though her affections sought constantly for some object upon which to repose themselves, for something to lavish that tenderness upon with which her heart was overflowing. Disappointed in her first choice—(which has been basely insinuated was her sister's husband—a tissue of falsehoods from beginning to end)—disappointed in those schemes of earthly happiness upon which her heart had once been set, she strove to forget all but her duty, and to love God alone: nevertheless, there were times when she could not help, situated as she was, desiring some respectable connexion and decent settlement in life; and it is believed that she received the attentions of several young men who professed to her honorable attachment, with the laudable object in view of obtaining such settlement. How different her fate would have been could she have been settled in life and tied to the duties of a wife and mother, we cannot now say, but the probability is she would have made a very respectable figure in society, and a much better wife than ordinary, owing to the natural docility of her disposition, her perfect habitual good nature, and forbearance and forgiveness. But the waywardness of her destiny prevented, and perhaps the providence of God, which sometimes ordains partial evils to promote some universal good, ordered it otherwise.

The religion of this ill fated girl, it will be seen by her letters, was a religion of feelings and frames.[1] Though there is no doubt it was sincere, yet it was of that unstable kind that is most apt to fail when most needed. She had engaged in it in a time of high excitement, and its existence was preserved—while it was preserved—by constant application of the means which created it: viz. by frequent attendance on those exciting meetings where highly wrought feeling and sometimes hysterical affection is often mistaken for devotion.

---

1. Reliance on extreme emotional states as the measure of one's spiritual life.

While there, there is no doubt she thought herself in the enjoyment of religion; and when out, the mind and spirits, by a natural reaction, would suffer a correspondent depression, and the same stimulus must be again resorted to. It will be observed that the style of the letters, which follow this slight sketch, varied materially after a years residence at Lowell, and were less frequent. Previous to this date, during a residence of more than a year in Dorchester, and the one year that succeeded in Lowell, religion seemed to be the chief subject of her correspondence; soon after which, it is evident the subject, for some reason or other, flagged: and her last letters, few and far between, do not even make mention of the subject. That there was a cause for this, no one can doubt. She could write of it when in the confusion of a boarding house, as she says, with "sixty boarders," and sometimes, nearly "all gabbling at once." But something has happened to damp her zeal now, or conscience whispers, "Thou that preachest to others, art thou a castaway?"

That she felt the want of a friend, that she desired one, is something so natural and proper that we cannot blame her for it. And that the warm tide of her affections sought for rest on some object was no fault of hers, but that they should have centred at last on a married man, was shocking indeed. That that man was her minister, the person who broke the sacramental bread, and presented the sacramental cup, was an aggravation of her crime, a heinous aggravation. Although it is to be presumed one of that sacred character might have more influence over the opinions of a person than any other; yet any attempts at familiarity ought to be doubly offensive in such, since it proves at once that he is a hypocrite.

As to the opinion of attachment on the part of S. M. Cornell towards her minister, we ground it on these facts. First, by her letters themselves; not merely because they shew a decline in religious zeal at the time when we believe it commenced, but from this circumstance:—that the name of Avery is never mentioned by her in any of them. She appears to speak with freedom of other persons, and other ministers; of Mr. Taylor, Oathman, Maffit and others; but his name she studiously avoids. She was not only three years at Lowell, the greater part of which time she sat under his daily and nightly ministrations, but she heard him at Great Falls and in other places in the neighborhood of Boston. Yet his name never escapes her pen. There must be some reason for this. As has been said, she seemed to contrive to be somewhere within the range of his preaching from the first of her acquaintance with him. Whether it was by her contrivance or his however, it is impossible for us to say, since she cannot tell, and he wont tell. How the intimacy commenced, and whether it was of a criminal nature previous to the Camp-Meeting at Thompson, we believe no one has taken upon themselves positively to say; but from what is related of the circumstances of their intimacy, every one can judge. The Parson, it is said, was a very polite man to females, frequently inviting some one of them to ride to a meeting or an evening lecture with him in his covered Carryall, and that he sometimes did the deceased the honor of riding with her. It will be recollected that at the Bristol examination Avery or some of his friends stated the fact that S. M. Cornell had lived a short time in his family, but that Mrs. Avery was not satisfied with her, and she had been dismissed.

At the time the sheriff passed through Lowell in pursuit of Avery, after his flight from justice, he learnt some very important particulars respecting this and others connected with it, and afterwards proposed laying it before the court upon his trial, but was told they were inadmissible, since it was not any particular act of impropriety in the prisoner's life, previous to the commission of the crime for which he stood indicted, but his general character which they wished to know, and which could alone in this case be considered as evidence. And as the sheriff was not prepared to prove that his *general* character, or that of any other preacher, was that of a rake, he of course kept it back: some of this found its way afterwards into the public papers of the day, and upon examination, the facts appear to be these.

First, that S. M. Cornell was a resident in the family of E. K. Avery about a week, and that during that time he used to come out of her room after ten o'clock at night; and that the family, on being questioned upon the subject, gave as a reason "that she was ill, and sent for him to come in and pray with her."

Secondly, that his wife, though habitually a mild, forbearing woman, on this occasion rose, and positively declared "she would not have the girl in the house any longer," when she went away.

Thirdly, that it was customary for him to be shut in his study with some young woman or other almost every day; sometimes several, in the course of the day. Very seldom any of these were seen by his wife; but that unfortunate woman was often seen with eyes red and swollen, as though she had recently been in tears; and though used to speak mildly, she never mentioned the name of S. M. Cornell but with evident resentment and bitterness of feeling, even after she had gone from there.

Fourthly, that he was in the habit of keeping very late hours; being out without his wife; and giving no satisfactory account of himself, not even to the family in the house, whose rest he often disturbed, by obliging them to sit up for him, as they did not feel safe to retire and leave the front door unfastened: that on one occasion, after returning from their own prayer meeting, at nine o'clock, (the time such meetings usually close,) and setting up* for Avery until after eleven, they retired, and he behaved with must unbecoming passion, beating and banging the door as though he would stave it in, and that the owner of the house hurried to let him in as quick as possible, and then retreated; when Avery entered, flung the door to, and snatching the key from the lock, carried it to his chamber. The master of the house followed him, and made him return the key. These things, together with others of an aggravating nature determined the family not to

---

* Nothing of this sort is credited by the author, or mentioned, without sufficient proof. Should it be necessary, those proofs can be coming forthwith.

reside any longer under the same roof, but having a chance to sell the house, they removed and left him in it. That it was not wholly on account of his late hours, so unbecoming in a clergyman, but on account of other things which they disliked; one of which was the frequent closetings with young women in the study, which stood at the head of the stairs and contained a bed; and was rather remote from the sitting room and lodging room of his wife, having to pass through the front entry and front room, and a passage way, to get to the kitchen where Mrs. Avery usually staid and lodged.

We do not place so much confidence in other things coming from a child, as they did, children being so prone to exaggerate and misrepresent; yet it appears the little boy of Avery, after having accompanied him on one of his rides, said on his return, "Pa kissed Sarah Maria Cornell on the road:" and that the feelings of the gentleman in the house were considerably tried upon observing at one time a wonderful alteration in the horse usually rode by him. His little boy accounted for it, by saying, that "the horse kicked his father, and he drove two spikes into the floor and tied his heels down, and kept him there two days without anything to eat or drink."* Of course, much was said respecting this man which was false: there is no one so base but may, after all, be slandered. For instance, the story of the mysterious and sudden death of his first wife must have been altogether false, for we cannot find that he ever had but one wife. There is enough of what justly belongs to this unhappy man, without any effort of imagination to add to it.

We have but one remark to make respecting the intimacy at the house, which is, that if an intrigue commenced at his own house, at that time; that if it was indeed true she used to send for him at that

---

* It appears that Avery is still famous for his treatment of horses. Few of his cloth would be seen to stop in the open street, get out, take his coat off, and beat a horse in the manner he has recently done in Bristol. "The merciful man is merciful to his beast" [paraphrase of Prov. 12:10—EDITOR.]. Though the story of the boy might not be correct, yet it was said the appearance of the horse warranted the conclusion that it was so.

hour of the night to come to her room to pray with her, *she courted destruction*, and might almost be said to deserve the fate it is supposed she met with at his hands. If, on the contrary, he stole into her room, without an invitation, the case might be a little different. That he was there, I suppose to be a fact. That she cherished an uncommon regard for him, criminal as that affection was in her case, was evinced, as Doct. Wilbour observed, "by the absence of resentful feelings." It was strange indeed, if she had suffered the injury she complained of, at the Camp Meeting, without manifesting any resentment afterwards, she should, on the contrary, uniformly speak of him and his family with tenderness, and above all things seem not to desire to expose him. It may be enquired, if this were the case, why did she leave that little bit of paper to direct, if she was missing to enquire of him? To that we answer, that our heavenly Father has implanted a something within us, that never fails to warn us of approaching danger: some call it "a presentiment of evil." But in her case there was something to fear exclusive of any resentment; that was, if her tale was true—if she had once had poison recommended her, and been warned by him who told her not to take it, neither to go to Bristol, nor to put herself in his power, but to have him come to her fairly and honourably, and settle it—if she had received this warning, she could not but have some fear. It was neither fair nor honourable in the first place, to ask a female to go to that cold, lonely place on a dark evening. She knew, probably, it was a fearful thing under such circumstances, or indeed under any, to go there to an assignation. The dark, deep waters of Mount Hope bay rolled below, and it would have been as easy to give one a plunge there, as to have poured down a dose of tansy oil.

That she had peculiar feelings of regard for this man may be inferred from the speech she made to Benjamin H. Saunders also. It does not appear there was any positive proof of any thing criminal in her conduct while at Lowell, by any testimony on the trial, if we except the testimony of the physician before named.

It seems S. M. Cornell was expelled from meeting while absent at a Camp Meeting on Cape Cod; and Avery tells, that he "advised her to go away while the process was going on against her:" but if the complaint was made against her previously, it was the height of impudence, to say no more of it, to suffer her to go to such a place, where the facilities for vice are so great. There cannot be, perhaps, exhibited, a greater proof of superstition, than the offer of this girl to make an acknowledgment to the meeting of what she, at the same time, solemnly declares herself to be innocent of, merely for the sake of being in church membership: for it was upon those conditions she offered it. As though to be out of the pale of the church was to be excluded from salvation. Her own words were said to be these, in a letter to Avery, where she gives a circumstantial and satisfactory account of her interviews with the physician:—"yet I will confess all, if I can only be continued in the church." Some suppose that the desire to be near a certain minister of that church was the great inducement, and that for his sake, or for the sake of being near him, she was willing to endure any disgrace, and would have signed any thing but her death warrant. There is one anecdote, which has been related to the writer of this, which proves she could not have been the abandoned creature represented previous to this. S. M. Cornell, at one of the places where she lived, worked in the employ of two brothers, partners in an establishment. Something had been said in their hearing about her not being prudent; and the oldest formed the resolution to find out how far her imprudence extended. He accordingly put himself repeatedly in her way, and at last insulted her with the declaration of his passion, which she resented firmly, and with some bitter reproaches. (They were both married men.) The older confided the affair to the younger, who felt piqued to try himself. He accordingly commenced a regular siege: but in the moment when he thought himself sure of success, met with a still more severe repulse than his brother. Upon comparing notes, they agreed it was only because they were married men; but as they felt somewhat in her power, concluded that it was not prudent to have her there. They accordingly gave her a hint her services were no longer wanted after which, being questioned with respect to her departure, said "she was rather too fond of young men:" though, as the gentleman said who related this and who, being in their employ, overheard the conversation between them, when they agreed to get rid of her, "he did not know what proof they had of her being fond of *young men,* except that she did not like *old ones.*"

Various anecdotes too have been related to the author respecting the charity, kindness of heart and gentle disposition of S. M. Cornell: but they would swell this volume beyond the bounds allotted to a work of this kind. Suffice it to say, that from all accounts, it appears her hand was ever open to the suffering poor, according to her slender means; that she was liberal to the society of which she was a member, and who did not disdain to receive of the pittance which her labour produced, towards promoting Methodism, whatever they may have thought of her character; that she was kind to the sick and afflicted; and retained a most affectionate regard towards her relatives, through the whole of her long absence from them.*

From Lowell she went to Great Falls, N. H. and here the same contradiction occurs with respect to what was said of her. A very decent and respectable young man who boarded with her the whole time of her residence there, has testified to the author, that he never knew of any thing being said or thought there, to her disadvantage;

---

* One woman, who has been very bitter against S. M. Cornell, and helped, it appears, to injure her at Lowell, gave as a reason to the author for thinking her bad, that she used to go up to Boston, sometimes, of a Saturday afternoon, on pretence of attending meetings "to hear good preaching," as she called it, and return Monday morning, "looking completely exhausted and worn out." I was amazed to hear such a reason given, knowing it must be great exertion, after tending three or four looms through the week to ride twenty-five miles Saturday evening, or afternoon, and then attend four or five meetings on the Sabbath and ride back again next morning. That, I have no doubt, was the fact; and if there is any female able to endure it, without feeling fatigue and exhaustion, and shewing it too, they must be hardy indeed.

that her conduct, and he saw her daily, was as becoming as that of any female he ever was acquainted with. Two respectable females too, who saw much of her during that time, testify to the same. Likewise at Taunton, where she next went, and where she had a place of considerable trust, having to keep the books of the weaving room, her character and conduct was respectable. So persons, who boarded and worked in the same room, with her, testified that "she was much made of," as he expressed it, and visited in very respectable company in the place.

While on a visit to her brother Rawson's in Woodstock, it was remarked that her conduct was strickly proper by the young men working with her brother, as well as by other members of the family; by customers at the shop and visiters at the house; and especially by Mr. Cornell, the Congregationalist clergyman, who, living quite near, was often in, and held frequent dialogues with her upon the subject of Methodist principles and discipline, for which she, of course, was a great stickler, and they observed, defended the cause with considerable skill.*

Previous to the fatal Camp Meeting, at Thompson, it seems her conduct there was without suspicion; and could she have rested content without another interview with Avery, it is probable this last final work of destruction might have been avoided. We have however the charity for her to believe, her intentions at this time to lead a new life were sincere, and that the interview, if planned by her, was only to obtain the letter of acknowledgment which it seems she had been influenced to write. That letter, she found was having a fatal effect upon her character, and what was of more consequence in her own view, was depriving her of her communion with the church; to secure which privilege she seems to have written it, expecting that

---

* A paper, containing a certificate from the Rev. Mr. Cornell is mislaid; but I recollect it testifies to the above fact, and to his opinion of her being a christian previous to the communication made to him by the family, after the Camp Meeting.

the command of scripture to forgive all who confess and ask forgiveness, would be literally obeyed. By the testimony of Benjamin
Saunders, who lived there and was in the habit of going to the post
office for her, it appears she corresponded with a minister in Bristol,
previous to the Camp Meeting; who the minister was admits not of
a doubt, especially as she requested him to keep it a secret, and by no
means to let her sister know. He recollected the circumstance by a
speech from her highly characteristic; on occasion of carrying one of
them to the office, said he, "I would not pay postage for a minister,
should think he was able to pay it himself." She answered, "I want to
help the minister all I can."

Through all the vicissitudes of life woman will be woman still. Of
the tenderness of woman's heart man can seldom form an idea; here
was an instance. There can be no doubt that this girl had great cause
of resentment towards the person she had been writing to, and very
possibly she wrote in harsh terms, requiring him to come, and bring
that letter of acknowledgment with him; but come to the trial, that
resentment could not even enable her to lay upon him the burthen
of paying the postage of a letter. No wonder she made so many
objections to naming a large sum as the settlement with her, which
the benevolent physician of Fall River recommended; she could not
endure the causing him inconvenience. That fatal tenderness too,
doubtless betrayed her at the camp ground; perhaps, and let us in
charity suppose it, she meant from henceforward to leave the path of
sin, and walk in that narrow one that leads to life, and trusted to her
resolutions to meet and part with him without any actions that virtue could condemn; the sight of him put all her good resolutions to
flight, and that beguileing tenderness again plunged her into misery
and irremediable distress. Her only road to safety would have been
in not seeing him at all. Gone was the look of cheerfulness she was
wont to wear; it was evident something pressed heavy on her heart.
The sense of her crime and the dread of its consequences at length
impelled her to yield to the solicitations of her affectionate sister,

and confide to her the humiliating cause of her grief and anxiety; that sister who had but one heart and mind with her husband, immediately sought counsel of him. What to do, or how to conduct themselves in such a strange case they scarce knew, and the brother finally resolved to ask counsel of his pastor, and subsequently of another friend, an attorney in the neighborhood. They advised her immediate removal into the State of Rhode-Island where Avery resided; and the brother feeling a delicacy about her remaining in the shop where his young men were, willingly acceded to the plan. She herself proposed to go to Fall River and work there in a factory while able to work, and until he should make some provision for her. Here she lived for about two months without reproach or suspicion, being perfectly correct, as every one supposed who saw her, in her conduct. In the respectable house where she boarded, and where there was a small family consisting of females, she was much beloved, having won their regard by the gentleness of her manners and the apparent amiability of her temper. To the daughter of this family in particular, she was in the habit of speaking with some confidence, when she showed those celebrated letters of different colors mentioned in the trial. This young lady remarked the pink and yellow ones appeared to be written by different hands, and that one looked like a lady's hand—and that S. M. Cornell answered, "but they are both written by one hand, by a gentleman in Bristol." This answer the witness was about to repeat in court, when she was stopped by the prisoner's counsel. She, (S. M. C.) told them several times that she was only waiting for some money she was expecting to receive, when she should leave Fall River. The flutter of spirits, which made her on the last day of her life more cheerful than usual may be easily accounted for. She came out of the mill early and changed her clothes, and then probably wrote that little strip of paper, "If I am missing enquire of Rev. E. K. Avery." Her habitual politeness never deserted her; even then while drinking tea, which was got early for her at her request, she said, "It is not very polite for me to be drinking tea here alone I know, before the rest, but I am in such a hurry," and turning to the oldest sister, she said as she went out, "I think I shall be back as soon now, as Lucy returns from the factory." Alas! she returned no more.

CHAPTER VI

Of the birth, parentage and early life of E. K. Avery, we know nothing, except that we are informed he was the son of a revolutionary soldier; if that be the case every one must rejoice he was spared the fate that threatened him in May, 1833. It would indeed be a kind of blot upon the history of the brave defenders of our soil, that one of their children should come to such an ignominious punishment, since it is the disgrace that constitutes the evil with men, though with God it is the sin. We are sorry to say we have so little account of the early years of a man who has made so much noise in the world; but from the time we can get any thing of his history, there seems to be something in almost every place that goes to prove him a bad tempered, daring and unprincipled man. As to his person and address we know nothing of them, but we conclude they must be extremely imposing, at least to the people he is among, since he seems to have been approbated and upheld by them on all occasions, except one; it seems he has been baffled once, and that was by a woman. Report says he studied the science of medicine, previous to his becoming a preacher, but whether he ever commenced practice previous to the 20th of December is not known; we have not heard of any of his performances since that date.

The first we know of E. K. Avery he is preaching in Duxbury, Mass. and from thence he removed to Scituate, Mass. in 1827, and took charge of the methodist society in that place. Among the people who now sat under his ministry was a maiden lady of about five and forty, who sustained a very high reputation for piety as well as for what they style *her gifts*. She was one of those active, useful women whose exertions were always called in, and always freely bestowed, wherever distress of mind or body required relief of any kind. The

EPHRAIM KINGSBURY AVERY
"He was a middle aged man, tall, and of very stout frame, and a face that might have passed for good looking, had not a certain iron look, a pair of very thick lips, and a most unpleasant stare of the eyes, have taken much from the agreeable."

young resorted to her for counsel, and the established christian for encouragement. She was as report said, not only a woman active in meetings and by the side of the sick and the dying, but what is extremely difficult, she supported on all occasions, a character for consistency that went far to make her labors successful. Of course the new minister soon discovered the real character of this lady; he not only heard her praises from every quarter, and witnessed her zeal, but he was also enabled to appreciate her excellence by personal acquaintance. She was evidently a woman of great spirit naturally, but so humble and subdued by the influence of real piety, that the very belief that this was the case could not fail in a sensible and candid person to increase respect for her. She was not a woman in dependant circumstances by any means, so there was no way to torture her or try her disposition that way; she was past the bloom of youth if not the meridian of life, and thought not of conquests—of rivalship and admiration; so there was no way to pique her in those—but humbled she must be, something must be done to try her temper until she proved herself mortal and no better than other folks. The new minister took a terrible dislike to her from the very first. He thought "the people put too much confidence in her," and averred "that he would see she was not made a goddess of." Whether her deportment was such as to shame some—a standing reproach to some others, who ought at least to be as consistent, or whether he thought others would rise higher if she fell, or whether she was one of those provoking women who have the faculty of reading characters at a glance, or from whatever cause we cannot say, but certain it is the Rev. E. K. Avery labored from the first of his going to Scituate to destroy this woman's good name, and thereby lessen her influence; at first the dislike was only vented in a few sneering remarks to her disadvantage, which she immediately heard of; those remarks being wholly unprovoked could not fail to create a degree of resentment in the object of them. During this frame it so chanced that the minister met her one evening at a prayer meeting, where he happened to call just after visiting the house of a parishioner who had lately lost his wife. In the course of the conversation he observed that "the husband (who was somewhat intemperate) will soon drown his sorrow." The lady upon returning to her lodgings, which was with a niece of this bereaved husband, repeated the remark; it was again repeated and created some little unpleasant feeling towards Avery for what they judged rather unfeeling and ill timed. His (Avery's) resentment against the woman was now at its height; he had something to seize upon, and although she blamed herself exceedingly for her own imprudence in mentioning his random speech, and with much humility asked his forgiveness, &c. yet it nothing mollified his ire. His hatred had now broke out into acts of hostility, and he commenced writing letters to various persons in Duxbury, and elsewhere, to try to get her expelled from the church, but all to no purpose. The woman, nothing daunted stood her ground manfully, and defied him to the proof of what he had asserted, viz. "that she had been guilty of lying and unchristian conduct, and exercising ungodly and unholy tempers," &c. The matter was before the parent church at Duxbury a long time, many letters passed between Avery and Mr. Mudge, as well as with others on the subject, but nothing could be proved against the woman, and it finally resulted in her coming off with honor, and with a certificate of her good standing. How they managed to retain Avery in his standing after his failing to substantiate his charges we cannot tell, but that was their business. The certificate made no mention of the recent trial and its result, but merely stated what they could not avoid stating, that she was in good standing in the meeting.—It was as follows:

Duxbury, April 12th, 1831.

This may certify, that Fanny Winsor, the bearer, is a member of the Methodist Episcopal Church in Duxbury, and is recommended as such by me the subscriber.

ENOCH MUDGE,
Minister in charge of said Church.

The church now located in Scituate has since had a recommendation of Miss Winsor, from that in Duxbury, signed by Daniel Fillmore, in behalf of the church in Duxbury. Report said that the friends of the lady were not satisfied with the proceedings of the meeting altogether, inasmuch as he had no censure passed on him for bringing those false charges, and that they wished her to go to law for redress, as they thought what he had said was calculated to injure not only her religious but moral character—and from a letter to Avery which she wrote in 1830, it appears she then threatened some such remedy, that is as we understand her language—The letter was as follows:

To the Rev. E. K. Avery.
Scituate, March 24th, 1830.

Sɪʀ—I address you from the purest motives of my heart, and under circumstances peculiarly aggravating. Your conduct towards me urges me to use my pen, which otherwise would have remained silent—this I do in my own defence. The charges against me, in writing from you to brother Mudge, he informs me are three. The first relates to a circumstance that happened three years ago this month,* respecting Benjamin James.† Now look at it candidly and see if you have experienced the right spirit. As respects the case, did I not confess at the time you brought up the accusation against me in the presence of my sister Susan—that I repeated your words inconsiderately, and was sorry, and said "any compensation you requested I was willing to make, even at your feet," and you would not** be reconciled—does not this bespeak that you would not forgive. Reflect—

---

* This persecution of Miss Winsor actually continued upwards of four years.
† The person who Avery mentioned "would drown his sorrow."
** Nothing can exhibit in stronger colors the ridiculous veneration in which they hold their ministers—at *his* feet truly!!

what does the gospel you profess to preach say: "If we offend seventy times, and repent and ask forgiveness we should be forgiven." Are not these the words of our Savior, whose image we ought to bear.

The second charge I think brother Mudge tells me, was "indulging unholy and ungodly temper." This I am confident was a false charge, as I know of no time whatever that I had any conversation with you, after that in my shop, in the presence of Susan, and I leave it to her if there was any thing of that manifested at that time—no, I was too much wounded in soul to indulge unholy temper. Your conversation towards me was like barbed arrows. What past God was a witness to, and his justice will be satisfied; for he judgeth impartially. I can say in his presence and his spirit accompanying me, I do feel clear of this charge. I know not what you have been informed by unholy people, that are plotters of mischief—they must see to it.

The rest I think was a charge of "talking to your disadvantage." This charge is as empty of truth as the other: The most I have said is this. When tale-bearers have brought to me, what they say, you have said, I have replied, "how can I hear such a man preach? that bears such a spirit? No I cannot! No nor will not—under existing circumstances." And I say so now, unless I view the subject differently. Sir the many times you have been to this place, you have not so much as changed a word with me on the subject since the time first mentioned, but if I am rightly informed, said behind my back, what you had ought to have said to my face. I am sure there has not been any time since the first awful moment but what I should have been glad to have settled the affair and buried it in oblivion.

But sad to relate, you seem to lay the axe at the root of my moral and religious character by this last move. This prompts me to take proper steps to vindicate my own cause, and clear up my character.

I am ready to settle it upon any consistent terms short of the law, that you are willing to. But if I hear no more of you, I shall put it into the hands of one authorized to do justice to you and me.

Take away our good name from among men, and you strike a death blow to all we hold dear in this life. Take away our good name from among our brothers and sisters in the Church, and then this world will be a barren wilderness. But one thing—no weapon formed against the child of God, can take away our name from the book of life.

I think defamation of character is an evil not to be overlooked or passed by unregarded—therefore I feel justified in putting myself in the way to have justice.*

FANNY WINDSOR.

Abraham Merrill, one of those who swore in court at Newport that he knew nothing against the character or temper of Avery, was knowing to all this transaction; we must suppose there were others who had like knowledge; and with how much truth or propriety could any one say they knew nothing against his moral character or his temper, that knew of such a diabolical persecution of an unoffending female, a defenceless woman, who probably was guilty of no real offence against him, or any one; and if she had been, who is to set examples of forbearance and forgiveness of injuries, if preachers of the gospel are not? If ministers were to commence a general dealing with all in their communions who exercise unholy tempers, it is presumed they would have their hands full, though in this case there would have been one innocent. There was testimony sufficient that she endured the bitter things so often repeated to her without manifesting any thing but sorrow. That she could not consent to hear him preach may be attributed to *principle* rather than temper.

The character of Avery for revengeful, angry feelings, may be gathered from the circumstances of the prosecution, by a brother clergyman. This was in the town of Saugus, Mass. and the circumstances

---

* "Justice," indeed—if this injured woman and the Rev. Thomas F. Norris have not been amply avenged by a righteous God, they never can be. "Surely there is a God that judgeth the earth" [paraphrase of Ps. 58:11—EDITOR.].

are related thus. The Congregational Society in that place were at that time destitute of a settled minister, and Avery, who was then stationed near over the Methodist one, offered to preach for them occasionally. The offer was politely accepted, and some little time after, a Mr. Norris, who was esteemed as a very amiable and pious man, and who was then preaching there to the Reformed Methodists, as they are called, (a sect of christians who have separated themselves from the others,) offered likewise. He too was accepted, and preached there much to the acceptance of the congregation, who were delighted with the unassuming piety and evangelical sentiments of Mr. Norris, and asked him to continue his labors among them, whenever opportunity offered. The next Sunday that Avery preached there he took for his text the passage in Job—"I also will declare mine opinion,"[1] and commenced an attack from the pulpit upon the character of his brother, whom he called a thief, and some other very bad things; and getting in a passion as he proceeded, went on to charge him with individual sins, which he undertook to particularize. A part of this discourse, as related, the writer has forgotten, but one was that he had been employed once in a glasshouse, and stole ware to furnish his own sideboard. His hearers who relate the story, remark, that "all this time his face was violent red, and he appeared to be in a great passion." The whole story was immediately related to Mr. Norris, who proceeded to put his character in the care of the law, and prosecuted Avery for defamation of character. It was tried, and Avery was found guilty, and sentenced to pay a fine, but he appealed, and it came to the second trial, when Avery appealed it, arresting judgment, and taking it out of court, by paying a sum of money, the amount of which we did not learn, but our informant says several hundred dollars. The Ecclesiastical Council, as they style themselves, then took him under their protection, and issued a manifesto declaring him entirely blameless, and clearing him of all censure.

---

1. Job 32:10, 17.

After the examination at Bristol, some of the history of this transaction got to Fall River, and a copy of the examination was forwarded
by some one to Rev. Mr. Norris, asking for the copy of the trial at
East-Cambridge. The amazement and indignation of Mr. Norris and
his friends, at finding the Merrills had sworn his character was unsul-
lied, &c. &c. together with the belief that the public ought to have
the facts, induced them to publish the following manifesto, which
was forwarded to Fall River, without the copy of the trial. That document he states was sent to the Governor of this State. No Governor
of this State has received it, and by what means it miscarried is not
known, but it is something that our public functionaries ought to
look into. Could it have been taken out of the mail between here and
East-Cambridge? If this book should fall into the hands of Mr. Norris, we hope he will himself see to it. The manifesto is as follows:

To The Public.

East Cambridge, Feb. 1833.

Fellow-Citizens,—I have frequently been solicited for a copy of
the trial and verdict in the action before the Supreme Court, at its
session in Cambridge, last winter, against E. K. Avery, but have hitherto denied.—Those solicitations becoming more numerous and
pressing, on seeing the strange testimony of the Messrs. Merrills, at
Avery's examination before justices Howe and Hale, I have permitted
some of the friends of justice to publish a few statements on the case,
with some animadversion on the evidence given by the Merrills, at
Avery's examination as published by L. Drury.

I feel no resentment towards E. K. Avery, and I write more in
sorrow than in anger—sheer necessity compelled me to shield
myself from his aspersions behind the strong arm of the law. And
what appears like an attempt to cover crime and screen the guilty,
by men in holy office, seems to render it proper the community
should have facts.

The verdict of the jury with their names, signed by the clerk of
the judiciary, has been forwarded to the Governor of Rhode-Island.

Avery has paid me one hundred and ninety dollars on the verdict, and paid his own costs, which probably amounted to as much more, as he summoned many witnesses. His friends offered in consideration of the abatement made him, to obtain his confession and retraction, to one of whom I returned the following written answer:—In respect to a confession from Mr. Avery, it would be highly dishonorable in me to extort one from him. The verdict of the jury fully shields me from all possible harm from the slander of his tongue, completely nullifying its utmost poison;—rather ought he to humble himself before that church of which he is a member and minister, upon whose escutcheon he has brought a stain, which the good conduct of a long life can never wipe off. I respectfully asked justice of Mr. Avery, and when tauntingly refused, I notified his superior, the Rev. Bishop Hedding, but obtained no redress, until I appealed to a jury of my countrymen. Should these facts be denied by responsible authority, the public shall have the trial and correspondence.

The following piece was prepared by a highly respectable member of the Middlesex bar, for and at the instance of several gentlemen of the counties of Suffolk, Middlesex, Worcester, &c. and is published by them;—some expressions of commendation of the writer are thereby accounted for.

Fellow-Citizens, your very obedient and humble servant,

Thomas F. Norris.

A pamphlet purporting to contain a report of the evidence given on the recent examination of the Rev. E. K. Avery, for the murder of Miss Cornell, is before the public, and much of it is clearly not legal evidence, and has no more to do with the question under examination than the history of Meg-Merilles.[2] This is our opinion, —others may view it differently. But not careing to quarrel about mere

---

2. A mythical Scottish gypsy, heroine of the ballad "Meg Merrilies" (1818) by John Keats.

matters of opinion, or rules of evidence, upon which even lawyers differ, let us notice a few facts.

On page 30 of this pamphlet the Rev. J. A. Merrill is made to swear that he had known Avery for about 11 years, and that as far as his moral, christian, and ministerial character was concerned, it is unspotted and unblemished. True it is on a cross examination, the Rev. gentleman is forced to confess (which he seems to have done reluctantly enough) that Avery, while at Saugus, got into a difficulty which resulted in a prosecution against him, a verdict against him, an arrest of judgement,—business settled, and an ecclesiastical council after the civil trial, acquitted Avery, and gave him a certificate. On page 45, Rev. A. D. Merrill is made to testify that he had heard the evidence of the Rev. J. A. Merrill, and concurred with him as to the unspotted and christian character of Avery, and that the prosecution against him at Saugus resulted in nothing to impeach his conduct.

Now to us it is to the last degree surprising that these Rev. gentlemen should have testified in this wise about Avery's character and conduct. The prosecution against Avery was a civil action, in which he was charged with publishing a most false, malicious and wicked slander against a peaceable, unoffending citizen and minister of religion. This charge was made in a variety of forms. Avery denied the truth of it, but notwithstanding this denial, a jury of his countrymen, after a long and labored defence, in which he was aided by the most eminent council, and a host of clerical and lay brethren, and the supposed sanctity of his own profession, declared on oath that he was *guilty*. This verdict was rendered upon the evidence of Avery's own religious and personal friends; and we have higher and better authority than the assertion of the Rev. Messrs. Merrills for saying it was a "most righteous verdict." It is true, that after this verdict was pronounced by the jury, the council for Mr. A. made a motion in arrest of judgment, on a point of special pleading, but even this ground was abandoned, and the matter settled before the time arrived for a hearing on the motion,—the object of it therefore was clearly to gain time.

The slander charged upon Mr. Avery was proved to be *wanton, malicious, false,* and wholly *unprovoked.* No circumstances appeared at the trial to justify, excuse, or even palliate this dastardly and wicked attack upon the character of one who was an utter stranger to Mr. Avery, and whose only offence was that of seceding from the great body of Episcopal Methodists and organizing an Independent Methodist church and society in Mr. Avery's neighborhood. The object of Mr. Avery seems clearly to have been to prostrate and ruin his opponent, and thereby to destroy the christian society he had laboured to unite and build up. The Rev. Messrs. Merrills were present at this *trial,* heard *the evidence,* and *knew the result;* and yet have taken upon themselves to swear that this prosecution "resulted in nothing to *impeach* his *conduct.*" Has it then come to this, that it is no stain upon a christian minister's moral character to be convicted of uttering falsehood and groundless calumny, and of propagating malicious slander against his brother? Is it not robbery to take from an innocent man the dearest and best of his earthly possessions? Is he, whose business it is to enforce the precepts of the peaceable religion of the Holy Jesus, and to preach charity and all long suffering, to gratify his own malignant passions, in traducing a brother and neighbour? Let these Reverend gentlemen look into that holy religion which they profess to teach, and see what St. Paul says of the slanderer, and what St. Peter says of the "man that bridleth not his tongue." Can that man's moral, christian and ministerial character be *truly* said to be unspotted and unblemished, when the records of our highest Judicial Tribunal show that he has been *accused* and *convicted* of an offence against the peace and laws of the land; against the rights of individuals, (an offence originating in malice,) and designed to blight the fair fame of an unoffending man? Let these Rev. gentlemen settle this question for themselves. Their consciences are in their own keeping. The slanderer, in the estimation of all good men, is no better than a robber or an assassin, and it will require something more than the ipsedixit[3] of two "holy men in holy office" to overrule public opinion, the verdict of a jury or the laws of the land. And before the bar of public opinion, we leave the Rev. gentlemen, and Mr. Avery also, to receive such judgment as their respective cases may deserve.

Thus far the manifesto. We will now go back to the history of Sarah Maria Cornell.

---

3. An unproved assertion.

CHAPTER VII

There is a wonderful mystery in the fact, if it be so, that this unfortunate girl should be constantly betraying herself to the Methodists, by confessions of guilt and self-accusations of sins of a most outrageous kind, while at the same time she was endeavoring to keep in the society, and be in fellowship with the members, and respected by them, striving as though her very salvation depended upon it. The trial has been published, and the evidence is before the public. Those who wished to make her appear a monster of wickedness, have continually said all that is possible to say against any individual, and said it as a certain preacher once said (when he was planning to abuse his neighbour from the pulpit)—from a place *where she cannot answer them back again.* It is however no more than fair that her letters should speak for her, and the author has been at the trouble to collect all of her correspondence that can be found, consisting of sixteen letters written to her mother and sister, all, except one, between the year 1819 and 1832. It will be seen by these that there is a period of more than a year when only one letter was written. This was the period immediately succeeding her troubles at Lowell, and may be accounted for by the agitation of mind which such a punishment or persecution, (call it which we please,) must have occasioned. It appears however that she was not entirely unmindful of her friends during this period, as by her last letter, dated March 10th, 1832, she speaks of a pamphlet sent to Mr. Rawson. And by a letter from him to her it appears the family received one on the 11th of Jan. 1831. Other letters, written at different times may have been lost or mislaid, but not by design. Her sister's family informed me that they were all of a like character, and, resembling her conversation, full of Methodism, and relating mostly to her religious feelings. The papers were all given up without reserve. Both hers and theirs were found among the few things at their house. The letters of her brother-in-law, Mr. Rawson, to her, are in themselves a complete refutation of any scandal propagated against him. They prove him to be what every one acquainted with him esteems him to be, a humble, plain dealing, and practical christian. They gave her excellent advice about her disposition to rove from place to place, and cautioned her of the danger, and expressed great satisfaction at her continued assurance of loving God and religion, and endeavoured faithfully to point out to her the necessity of giving herself up wholly in a life of good works, and not to rest in a mere profession. There is also among her papers letters from some of her Methodist sisters, expressing fellowship and christian affection.

One of these letters, written in 1827, from, as it appears, a pious and quite intelligent young lady, styling her worthy sister, &c. struck me very forcibly as being the year after what they term her "disgraceful expulsion from the meeting at Smithfield." It appears they had lived together, and been for some time in habits of intimacy, and expresses great desire to have Sarah Maria follow her to the place where she then was. One from another sister, dated 1829, also addresses her as a "worthy sister," and feelingly asks an interest in her prayers, and dwells upon the seasons of religious enjoyment they have had together. One was directed to her at Dorchester and another Lowell. Her letters here follow, copied verbatim. The originals are now in the hands of the author of this book, and can be seen by any one who has the curiosity to see them in her own hand writing. The first is dated at Norwich. (One letter, No. 1, is omitted simply because it is a child's letter, written at 12 year's old.)

Letter No. 2

Norwich, May 6th, 1819.

MY DEAR SISTER—Having an opportunity to send directly to you I thought I could not let it pass without improving it. My sister, the time is coming when we shall prize time better than we do now, when we shall improve every moment of the short space allotted us. I have this afternoon received the parting hand of our dear cousin Harriet, aunt Lathrop's eldest daughter, she lately married Mr. Winslow a missionary—and is to embark for Ceylon, never expecting to see her beloved parents in this world; but she is an example of christian piety, she has left her native home to go to instruct the ignorant Heathen who sit in darkness worshiping wood and stone* and know not the God that made them. Let us inquire my dear sister who made us to differ? We have the Bible and are taught to read it. Let it be our daily prayer that God would send more missionaries to the heathen, to spread the gospel to those who know it not.

I am learning the Tailors trade, I have been here seven months, and expect to stay 17 more. I hope when my time is out I shall come and see you, I expected to have come last fall—but was disappointed. Mother is well and sends her love to you, likewise Granma—Uncles, aunts, and cousins. But where is our beloved brother, I have not seen or heard from him these twelve months, May God Almighty help guide and direct him and us, and bring us safe to heaven. Give my best love to all my friends, and you must write me as soon as you receive this—either by public or private conveyance. We have been so long separated that we should not know each other by sight, but surely we might have the pleasure of corresponding. You must excuse this scrawling and I hope the next will be better. Adieu my dear sister.

I remain your ever affectionate and loving sister,

SALLY MARIA CORNELL.

---

* The late Mrs. Winslow, wife of the Missionary of that name, who died lately at Ceylon, was first cousin to S. M. Cornell.

Letter No. 3

Norwich, August 26th, 1820.

MY DEAR SISTER—I received your letter about three weeks since but have not had time to answer it till now, being very much hurried in the shop. Mother has had two letters from our brother since I wrote you last. He was then in Natches, but has gone to Fort Gibson, and says he has very good business, and shall be at home next summer if possible.

Your sister M. with all your friends rejoice at the change the Lord has wrought in your heart. O that he would condescend to visit your poor sisters heart also. There has been quite a revival here, about twenty I believe is going to join the Church next Sabbath. Our cousin Leffingwell aunt Lathrop's youngest son is very serious, a year since they could hardly persuade him to go to meeting on the Sabbath, but now he is one of the Sabbath school teachers a young lady who has had a consumption for about a year, dropt away suddenly yesterday. When we see one and another of our friends dropping into eternity it ought to remind us, that this is not our home or abiding place. It naturally leads us to enquire was they prepared to meet death and the judgment? The young lady I mentioned that died yesterday was resigned and took leave of all her friends, and said she hoped to meet them all in a better world, she said she could bid defiance to death, and meet Jesus with a smile. O that my feelings were like hers, but alas my heart is hard, and I am as prone to sin as the sparks that fly upward. Oh my sister pray for me, that God in his infinite mercy pour the sweet refreshings of his grace on my soul.

I have almost finished my trade, my time will be out in October, and mother is making preparations for our coming to Providence this fall. Oh shall I behold the face of my beloved sister which I have never seen—or have no recollection of.

Although we are strangers we ought not to be deprived of the privilege of writing to each other. Only think we are only forty-five miles apart and we dont hear from each other more than once or twice a year—and our cousin Harriet is three or four thousand miles from her parents and they have heard from her four or five times, she is well and has never regretted devoting her life to a missionary cause, she says if she is a means of helping bring the heathen out of idolatry she shall be doubly rewarded.

Mother Grandma Aunts and cousins send their love to you, and would be very happy to receive a visit from you. Give my love to all my friends in Providence. Oh that you and they may be useful in this world, and happy in the world to come is the prayer of your affectionate sister.

Sally Maria Cornell.

Letter No. 4

Bozraville, May 3d, 1821.

My dear sister—I with pleasure resume my pen to inform you of my pleasant and happy situation. I have been at Deacon Abels all winter and have just been able to pay my board, I am now situated in a pleasant village near the factory, and between the town of Bozrah and Goshen, four miles from mother, and three from the meeting-house, we have meetings in the factory every Sabbath, and when it is unpleasant I attend. I am the only Tailoress for two miles each way, you may of course conclude I shall be somewhat hurried with work. I wish you were here. I desire to be thankful to God for placing me in so pleasant a situation.

The solemn bell has just summoned another fellow-mortal into eternity but what is to be his fate in another world God only knows. It is just four weeks since death entered Deacon Abel's family and deprived them of a servant—a tall stout robust negro whom they had brought up from the age of two years, twenty years he lived with them, and never associated with any but respectable people, as there was but one other negro in the place. Deacon Abel's family took his death very hard, he was in the vigor of health, often boasting of his strength—but when he came to be laid on a bed of sickness and the cold hand of death was upon him all his strength could not save him. he had just finished his years work, and engaged for another year, and wanted one week for relaxation, and two weeks from the day that his year was up he was carried to his grave. the family did not consider him dangerous[1] until just before he died, but he was imprest with the idea he should not recover and regretted that his life had not been better, and thought if it should please a just God to spare him he should live a different one. it is not for us to say whether he is happy or miserable in another world, but his death has very solemnly impressed my mind. Sometimes I think *why am I spared* perhaps it is to commit more sin, perhaps for some usefulness. sometimes I think I am no worse than others what have I to fear but God says be ye also ready for ye know not what hour your Lord will come.[2] How will ye escape if ye neglect so great salvation.[3] Yesterday I heard a discourse from these words "Why halt ye between two opinions, choose you this day whom you will serve, if the Lord be God serve him, if Baal then serve him."[4] I have thought seriously about this text.

You will perceive by the date of this letter that it is my birth day. Nineteen years has rolled round my head and what have I done for God? If I were summoned before his judgment bar could I answer with a clear conscience to having performed my duty? I fear I could not.

I have resolved this year, to leave the world and all its glittering toys, and devote the rest of my life to the service of God. I have searched this world for happiness, but alas I have searched in vain; it is all a mere show—a broken cistern that can hold no water.

In your last letter I recollect you harbored the idea that I was offended with you. Far be it from me to be offended with my sister—you took my letter very differently from what I intended it. I received a letter a few days ago from James; he has changed his situ-

---

1. In danger.   2. Matt. 24:44.   3. Heb. 2:3.   4. 1 Kings 18:21.

ation and will not come to Connecticut this year, therefore I shall give up the idea of visiting you this summer—a year from this time if God permits, I shall anticipate the pleasure of visiting you, but it is very uncertain. I had forgotten to mention I am boarding with one of the best of families, a pious woman and steady man. Please direct your letters to Bozrahville, to the care of David L. Dodge; there is a post office here and it will be more convenient for me to get the letters; write immediately on the receipt of this. I am so far from mother that it will not be convenient for her to write any more. Give my love to uncles, aunts, cousins, and all who inquire after your affectionate sister

MARIA.

P. S. Don't exhibit this scribbling to any one.

S. M. C.

Letter No. 5

Killingly, May 20th, 1822.

DEAR SISTER—I received a letter from you soon after I came to this place, in which you murmured at my coming to the factory to work; but I do not consider myself bound to go into all sorts of company because I live near them. I never kept any but good company yet, and if I get into bad it is owing to ignorance.

I have been away from home now about one year, and have found as many friends as among my own family connexions. I have learned in whatsoever situation I am in to be content, though I have not been so contented here, being far from any friend or connexion.

You wrote me you thought I had better return to Norwich as soon as possible, and that you should not come to Killingly as long as I staid at this factory. You must remember that your *pride must have a fall.* I am not too proud to get a living in any situation in which it pleases God to place me. Remember that you have expressed a humble hope in God, and bear the christian name; learn then to imitate the example of Him whose name you bear, and never let it be said of you that you were too proud to follow your Saviour's steps—who was meek and lowly and went about doing good— suffering the scoffs and indignation of wicked men, and finally spilled his precious blood that you might be saved.

I do not expect to find the society here that I did in Bozrahville. I have got some acquainted with Mr. A—'s family and like them very well. I miss Mr. Dodge and his family, and some other friends I left there; shall never enjoy myself so well in any other place as I did there. No my dear sister, there is no revival of religion here, and I have no class in the Sunday school here, and it cannot be expected I can enjoy myself so well.

If you do not come to Killingly until I go to Norwich you may not come this year, and I assure you I will never come to Providence first.

I had a letter from our dear brother a few weeks since; he is in New-Orleans, and he writes that he don't know when he shall return to Connecticut. I should be pleased could we all meet once more, but I don't expect we ever shall. My dear sister, may God be your guide—and may his holy spirit refresh and comfort you, and that we may both meet in heaven is the prayer of your affectionate sister,

SALLY MARIA CORNELL.

Letter No. 6
Slatersville, (Smithfield,) 1824.

MY DEAR BROTHER AND SISTER—Almost two years has elapsed since I have written a letter or hardly a line to any one, and I scarce know what to say to my dear parent—but through the goodness of Divine Providence I am alive and in a comfortable state of health. I enjoy all the necessaries of life and many of its enjoyments. I can truly say my dear mother, that the year past has been the happiest of my life. I have lived in this village almost nineteen months, and have boarded in a very respectable family. My employment has been weaving on water looms; my wages have not been very great, yet they have been enough to procure a comfortable living, with economy and prudence. I feel as though I had done with the trifling vanities of this world—I find there is no enjoyment in them and they have almost been my ruin.

While I am writing perhaps you have long since forgotten you have a daughter Maria—but stop dear mother, I am still your daughter and Lucretia's only sister. God in mercy has shown me the depravity of my own wicked heart—and has I humbly trust, called me back from whence I had wandered. Although I had professed religion, and have turned back to the beggarly elements of the world, and brought reproach upon the cause of God—and have caused Jesus to open his wounds afresh, and have put him to an open shame—and have followed him like Peter afar off—and even denied that I ever knew him.[5] When I look back upon my past life it looks dreary, and I feel like a mourner alone on the wide world without one friend to cheer me through this gloomy vale—but when I look forward it bears another aspect. I have been made to rejoice in the hope of the glory of God. I feel that I have an evidence within my own soul that God has forgiven me, and I have an unshaken trust in God that I would not part with for ten thousand worlds. I find there is nothing in this vain world capable of satisfying the desires of the immortal mind. But the religion of Jesus is a fountain from whence joys of the most exalted kind will for ever flow. I have enjoyed some precious seasons since I have been in this place. Though destitute of any natural friends,[6] yet God has raised up many christian friends of different orders—all united heart and hand, bound to one home.

We have a house for worship and have preaching every Sabbath.

Sister Lucretia, by the best information I can obtain, since I saw you last you have become a wife and a mother. I want to see the dear little babe; I hope the cares of a married life has not separated your

---

5. See Matt. 26:57–75.　6. Relatives.

heart from God. I believe there is something in religion that is durable; it is worth seeking and worth enjoying I feel as though I could enjoy myself in this life while blest with the presence of Jesus, I have found that a form of godliness will never make me happy but I can praise God for the enjoyment of every day's Religion—it is that which will do to live by—and will prepare us for a dying hour.

May God bless you and your companion, and if I never meet you in this world, may we be prepared to spend a never ending eternity together in the bright mansions of glory. I want to see Mother and if any of you desire to see me—write and let me know and I will try to come and spend a few days with you before long—but whether I ever see you again or not, I want you should forgive me* and bury what is past in oblivion and I hope my future good conduct may reward you. I heard that brother James past through Providence, if he is with you give my love to him. I should like to see him but never expect to.

Farewell in haste yours
Maria S. Cornell.†

The kind of self-accusation contained in the second paragraph of this letter is very common among enthusiastic people when making their confessions of sin. I have heard men of integrity—and young innocent girls, get up in meeting and roundly accuse themselves of crimes—the least of which, if any other had accused them of, would have been a mortal offence. Some very sensible and intelligent persons have done this in reference to the spirituality of the law of God which makes, they say, "an angry word murder, and a wanton look adultery."[7] (Vide Matthew v. 28.) We ought however to deprecate the custom, as it is most generally made a very bad use of.

---

* Alluding to the affair at Mr. Richmond's and Mr. Hodges' [in which Cornell had been accused of shoplifting.—EDITOR.].

† When baptized by the Methodists, she took the name of Maria, but having been accused of changing her name, afterwards resumed the old manner of signing it.

---

7. Matt. 5:21–22, 27–28.

Letter No. 7

Slatersville, Sept. 6th, 1825.

Friday evening half past seven o'clock.

My dear sister just before the bell rung, I heard of an opportunity to send to Killingly tomorrow by Frederic Dean, who is going to carry his sister home. I was truly pleased with my visit at your house, to see you thus happily situated, with your family around you. I hope dear sister you will never have cause to grieve again on my account if I know my own heart, I desire to live so that none may reproach me, or say "what doest thou more than others?"[8] I have enjoyed some precious seasons, since I returned from camp meeting. Sometimes when I think of leaving Slatersville, it strikes a dread upon me. Can I ever leave this delightful spot, where I have enjoyed so many delightful seasons and privileges, it seems to be a place highly favoured by God. Elder Tailor preaches here half the time, he is a powerful preacher, reformation follows him, wherever he goes he draws about as many hearers as ever John N. Maffitt did, some came eighteen miles last Sabbath to hear him. I wish you would send me word whether James has gone or not. Give my love to mother, tell her there is no small darning needles in the store. William and Eliza Nanscaven is coming up christmas, I shall send mothers gloves by them. Remember me to Mr. Rawson, I can never be thankful enough to him for all his kindness to me. It is growing late and I must bid you farewell in haste, your affectionate sister.

SALLY MARIA CORNELL.

Letter No. 8

Slatersville, Dec. 18th, 1825.

To Mrs. Lucretia Cornell.

MY DEAR MOTHER—Once more I take my pen in hand to answer a letter which I received from you not long since, in which you informed me my brother was gone. William and Eliza Nanscaven is

---

8. Matt. 5:47.

going to Killingly next Saturday. I have been making calculations all
the fall of coming up with them, but I am disappointed I have lost so
much time, I have been out sick a week—and last Saturday I went to
Douglass to quarterly meeting—and Mr. Osterhold is not very willing I should stay out of the factory so soon again.

Dear Mother we have good times in Slatersville Meeting almost
every evening. There are still many inquiring the way to Zion, I have
seen this summer and fall past nearly 30 persons own Jesus by following him down into the water in the ordinance of baptism, I have seen
the aged, the middle aged, and the blooming youth, the drunkard the
profane and the profligate all bow to the sceptre of King Jesus and say
though I have been a great sinner I have found a great Saviour.

I have reason to praise God that ever I was redeemed by the blood
of Christ, and that I was made an heir as I humbly trust of the grace
of God. Join with me my dear parent in supplication at the throne of
grace that I may be kept in the way—that I may never return to the
beggarly elements of this vain world—but that I may adorn the profession I have made by a well ordered life and conversation. I expect
the Lord willing to spend my days in Slatersville* I dont want great
riches nor honours—but a humble plain decent and comfortable living will suit me best.

You mentioned you had some yarn you would let me have I
should have been very glad of it, if I could have got it—but they
bought some at the store, and I have got as much as I need at present.
I wish you would send me word by William if you have heard from
James, and where he is, that I may know where to write to. I received
a letter since I saw you from our good friend David Austin. Remember me to brother and sister Rawson. I think my friends never
seemed so near to me as they do at present. I want to see little Edward
very much. I expect if it is good sleighing in February to come and

---

* She left the factory in Slatersville in consequence of its burning down, and went to the
Branch Factory. Not being contented there, she removed after some little time to Mendon
Mills.

spend one night with you if nothing prevents. I have no more to write but remain your affectionate child.

SALLY M. CORNELL.

P.S. Excuse the blots I am in a great hurry.

Letter No. 9

Mendon Mills, 1 mile from Slatersville, August 6th, 1826.

MY DEAR MOTHER—I left the Branch Factory, and came to this place about three weeks since, and am weaving blue Sattinet. The water was so low and filling[9] so scarce, the weavers could not do much during the warm weather. The factory that is rebuilding at Slatersville is going up slowly. I anticipate much in returning to that delightful village and seeing it assume once more that lustre that shone so brilliantly.

I received a letter from you some weeks since, in which you thought you should not probably be at home until September. I think some of going to Camp Meeting at Woodstock where I went last year, and if I thought you and Lucretia would be at home I should come that way and spend one night with you. Camp Meeting is appointed the 29th of this month and holds four days, some of my Methodist friends from the village will probably go with me. I am boarding at a very still boarding house of about twenty boarders. I enjoy myself very well most of the time. I meet my brethren and friends at the village about once a week.

I think much of my dear brother and sister Rawson in the afflicting dispensation with which God has been pleased to visit them. May they bear it with christian fortitude, and that it may be sanctified to their eternal good is the prayer of their sister. Give my best love to them and I should be much gratified to receive a letter from them.

My own dear brother—where is he? I have sat down several times with the intention of writing to him, but my heart has failed me I

---

9. The woof of yarn or thread laced through the vertical warp in a woven fabric.

know not what to say. If you are still at uncle M's remember me to them and tell them I am still enjoying that happiness which is the privilege of God's dear children to enjoy—feeling a desire to spend the remainder of my days in the service of Him who has done so much for me. Tell cousin Polly and my other friends in Providence, that I hope they will forget and forgive what is past, and I should feel very happy to receive a letter from them. I wish you would let me know when you expect to return to Killingly. In haste your affectionate daughter.

MARIA CORNELL.

William and Eliza Nanscaven are going to Camp Meeting.

[During the period between this letter and the preceding one, S. M. C. made a visit to her friends in Providence, meeting by appointment with her mother in that place. Whether the factory in Slatersville to which she proposed to return, had gone into operation at this period, we do not know, but when she left Millville or Mendon mills it had not, and a young lady of that village had agreed to go with her to work at Dedham. The difference between weaving cotton and woollen cloth is very great, and few persons accustomed to work on the former like the latter. No other reason is known for the removal. The following letter was written to her mother and friends about six months after parting with her, at Providence.]

Letter No. 10

Dorchester, Mass. Sept. 25th, 1827.

MY DEAR MOTHER BROTHER AND SISTER—After waiting nearly six months for a letter in vain, I take up my pen to address those of my dear friends who are near and dear to me by the ties of nature. After leaving you at Providence I came in the stage to Dedham where I found the young lady as I expected from Slatersville. I went to weaving the next day at Dedham, where I staid about four weeks. I immediately wrote, as I supposed before you left Providence, but as I have received no answer I have reason to suppose you have never received it. There was no meeting at Dedham that I wished to attend, and I had to board where there was sixty boarders, and after four weeks I removed to this place, which is about four or five miles from the city of Boston. It is a pleasant thick settled village. There is one Unitarian, two Congregational or Calvinistic, and one Methodist meeting in this place. I have spent some time in Boston of late. I frequently attend meeting there, at the Bromfield Lane Chapel. The Rev. Mr. Maffitt and Merrill are stationed preachers there. Mr. Sias preaches here occasionally and I have every thing to make me contented and happy but natural connexions, I have been expecting all summer to visit you this month on my tour to Ashford Camp Meeting—and had engaged a passage in the stage, but I found it would be so expensive—and I could stay so short a time—that I concluded to give it up—and go to Lunenburg with my Boston brethren. We started for that place August 28th, forty in number, in six private carriages. It is a distance of fifty miles. We had good weather all the time. Between 20 and 30 ministers were present, and about five thousand people. Nearly forty persons professed to have past from death unto life. Friday which was the last day of the meeting between five and six hundred professors partook of the symbols of our Saviour's dying love. It was a circle formed within the tents. The scene was truly affecting—it will no doubt be remembered by hundreds through time and eternity.

I reside about half a mile from Mr. Oathman's father's that used to preach in Providence, he is frequently here and preaches. The good people of Dorchester have ever treated me with the greatest respect. But it is uncertain whether I spend the winter here or in Boston. I have had several opportunities to work at my trade there, in shops where the tailors hire fifteen or twenty girls to make coats and nothing else. I should like to come and work a month with Mr. Rawson if I could—but I cannot this winter, it would cost all of eight dollars to go to Killingly, and back again—and my health has been very poor this summer, and I have not been able to work all the time, but through the goodness of God I am comfortable— though much has been said, and I have suffered very much from false reports in time past.

I enjoy myself as well as I could expect among strangers, as I have never seen but three faces since I left Pawtucket* that I ever recollect of seeing before, viz. Mr. Maffitt, Mr. Oathman and Lydia Knight, from Smithfield. After all that is past I have been sustained and upheld by the mighty power of God, and still retain a respectable standing in the Methodist Episcopal Church—and enjoy a comfortable degree of the presence of God. Dear Mother if you have any regard for me do write if it is only two lines, and direct to Maria Cornell, Milton Mass. as the Post Office in Dorchester is several miles from me, and I should not get it in some time. Milton office is only across a bridge—I shall come and see you another summer if I live and do well.

Yours affectionately,
Maria Cornell.

Letter No. 11
Dorchester Village, March 2d. 1828.

To Mrs. Cornell.   *Sunday noon.*

My dear Mother—Once more I take up my pen to write a few lines to my parents, as nearly six months have again elapsed since I have heard from you. I dont hardly feel reconciled to think so many connections and friends as I have in Connecticut and Rhode Island, that I cannot hear from any of them oftener than once in six or seven months. Sometimes I think they have lately forgotten me, but I have no reason to complain, I have cause to be thankfull that it is as well with me as it is. I am tolerably well and in good spirits—though I have never been well enough to work one whole month since I have been here. My work has been very hard the winter past, and I have got almost beat out, I have been weaving on four looms at the rate of

---

* She had stopped in Pawtucket on her way down, to see some connections residing there.

120 or 30 yds. per day, at 1 half cent per yard, my board and other expences are considerable here, I feel a good deal attached to the people in this place, being surrounded by some very dear friends, I have a very pleasant boarding house, and every thing around me to make me contented and happy. It is about one year since I have seen any of you, though to me I trust it has not been altogether an unprofitable one; my enjoyment has been great—and my privileges very many. I long to see my brother and sister, and the dear little babe, and I have been seriously thinking of visiting Connecticut the summer coming, if Mr. Rawson expects to stay in Killingly another year and it should be agreeable to you all, I think I shall come and spend a week with you some time in the course of the summer. You will please to let me know before the first of April, as I want to know how to make my arrangements.

There has been several shocking cases of suicide within a few months here, one of which a man about 30 cut his throat yesterday a few rods from me, he is to be buried this afternoon, he was intoxicated. I have not yet felt as though I could see him it brings so fresh to mind the murder at Smithfield I felt as though I had rather not see him.

About the first of February a young man shot himself before my face and eyes, I was looking out of my window. He tied himself to a tree and placed the gun to his breast, and before any one could get to him, made way with himself. A girl belonging to this establishment threw herself into the river, after remaining two days in the water she was found, the most awful sight I ever beheld. How short and uncertain life is, it vanishes like the early cloud and the morning dew. It is time to go to meeting and I must close. Give my last love to Grindall and Lucretia, and tell dear little Edward aunt Maria wants to see him very much.

Adieu, I am your affectionate though unworthy child,

MARIA CORNELL.

Letter No. 12

Dorchester Village, 28th 1828.

I received yours dated March 18th and was glad to hear you was all well, my health is pretty good at present, you mention you expect to visit Norwich this summer, I wish it was so that I could come and go with you, but I do not think it will be possible, as I have lately given five dollars for the purpose of erecting a new Methodist meeting house in this town, which is to be built by subscription, and you had better make your calculations to go to Norwich early as you can as you will probably stay some time. I expect to be in Killingly somewhere about the 20th of August and I should be sadly disappointed if you was gone.

You will please present my best respects to uncle and aunt Lathrop tell them that I long to see them, and if it is my aunt's wish to see me I should be pleased to have her write by you. I desire likewise to be remembered to the Rev. David Austin, tell him I wrote to him some months since, but as he has not answered my letter I conclude he has forgotten or wishes to forget me. I likewise desire to be remembered to Deacon Abel and his wife, Mr. Huntington and his wife, and particularly to Lucy Abel, and all others who enquire after your daughter Maria. I wish you to write me immediately on your return to Norwich, and if you cannot be at home the time I have set, you must let me know.

Adieu, with my best love to all—your affectionate daughter,

MARIA CORNELL.

Letter No. 13

Lowell, Jan. 11th 1829.

To Mrs. Cornell.

MY DEAR MOTHER—It seems a long time since I have heard from you, and I almost begin to think you have forgotten me or you would have written before this. I have written two letters and sent two papers since I have resided in this place, and not received a line from any of you. I hope you will consider I am a stranger in a strange land,[10] exposed to sickness and death. Last Saturday night about twelve o'clock I was called a second time to witness a five story factory with all its machinery enveloped in flames. It was a bitter cold night and with great difficulty they made out to save the others which stood on each side—there were five of the same bigness in the yard. The middle one caught at the furnace and in less than three hours it was burned to the ground. I expected to have seen the whole thirteen, with the whole Corporation swept by the flames. But through the goodness of that God who rules the elements— although the air was keen and cold—it was still as in midsummer. The damage is great, but the distress is nothing to what it was in Slatersville—as each factory supported itself. No one was personally injured. It was my lot to remove on the other side the river, about half a mile distant.

I feel measurely happy and contented, but do long to return to Connecticut to see my friends—but when I shall is unknown at present—think I shall never set any time to come, but hope I shall next summer if health and strength permits.

I want you should write as soon as you receive this—if you never do again—and inform me how they all do at Norwich. My best respects to my brother and sister—I hope they are doing well—and the children, with the sincerest affection I am your unworthy daughter,

MARIA CORNELL.

Letter No. 14
Sabbath morning, Lowell, May 3d, 1829.

Mrs. Lucretia Rawson.

DEAR SISTER AND FRIENDS—I take up my pen once more to inform you, that through the mercy and goodness of God, I am spared to see one more anniversary of my birth. Twenty-seven years

---

10. Exod. 2:22.

of my short life has rolled on to eternity, and I am still on the shores of time, a probationer of hope, and enjoy the day and the means of grace. More than two years have past by since I have seen any of you, or indeed scarce seen one individual that I ever saw before, but still I am contented and happy. I am surrounded by many dear friends who are near and dear by the ties of friendship and grace, and I feel much attached to the place and people here, and the religious privileges I enjoy are much greater than they have ever been before. But still I often look back and think of my natural connections in Connecticut and Rhode-Island, and long to be with you. I have been thinking of coming to see you for two summers—I feel a greater desire to see you now than I ever have done. I begin to think if I do not come to Killingly this summer I never shall. I received a letter from mother about four months since in which she mentioned she thought I was a moving planet, but I would tell my dear mother that I do not think I have moved much for two years past. I staid in Dorchester more than a year, and it will be a year the 17th of this month since I came to Lowell—and more than all this tell mother she must remember that I am connected with a people that do not believe in tarrying in any one place longer than a year or two years at most at any one time—and I am with them in sentiment believing with the Apostle that we should be as strangers and pilgrims having here no continuing city or abiding place, but seek one to come.[11]

With regard to my views and feelings respecting religion, they are the same as they have been for two years past. I was a great sinner but I found a great Saviour. Tis true I had made a formal profession of religion, but when I was brought to see and feel the necessity of being deeply devoted to God, my views and feelings were vastly altered. I am satisfied for one that a form of godliness will never prepare a soul for the enjoyment of heaven. For "great is the mystery of godliness. God manifest in the flesh—justified in the spirit—

---

11. Heb. 11:13, 13:14.

believed on in the world, and received up into glory."[12] Perhaps my friends may think strange that I chose a people different in their views and opinions from that which any of my friends have embraced. But let me tell you my dear sister that the Methodists are my people—with them by the grace of God I was spiritually born—with them I have tried to live, and if ever permitted to enjoy the happiness of the blest in heaven shall probably praise God to all eternity. I see my beloved sister a fulness in the Saviour, and I believe it is the privilege of the child of God to enjoy all the depths of humble love.

It seems inconsistent to me for the profest followers of the meek and lowly Jesus, who have said by their profession that they have bid farewell to the world to follow its customs and fashions. It has appeared to me some time that it was good for the proud heart to be adorned with the modest livery of God's dear children, and to have a daily evidence that our witness is in heaven and our record on high. The bell rings for meeting and I must draw my letter to a close. If nothing more than what I know of prevents I shall be in Killingly some time between the middle of August and first of Sept. I do not know why you or Mr. Rawson have not written to me. I want one of you to answer this previous to the first of June and let me know what your wishes are, and I shall act accordingly. I am affectionately your sister,

MARIA CORNELL.

P.S. I am obliged to write where there are 30 or 40 boarders a gabbling—so excuse mistakes.

Letter No. 15

Lowell, Jan. 17th, 1830.

To Mrs. Cornell.

MY DEAR MOTHER—After waiting for more than eight long months for an answer to a letter that I wrote you last spring, I once more take

---

12. 1 Tim. 3:16.

up my pen to address you. You wrote me then you were going to visit your friends at Norwich, and that you would write me immediately on your return, but as I have never received a line from that time, I have concluded that you were there or were sick or dead, for it appears to me if you were in the land of the living and possess a parent's feelings you would have written before this. When I last wrote to you that if the Lord spared my life and health I should visit Connecticut in August last past. A long time I waited for your return from Norwich, thinking you would write and let me know, but at length concluded it was neither your wish nor that of my brother and sister that I should visit Killingly—but enough of this—I will cease to trouble your minds with such painful feelings. Not a day has rolled over my head since I left you but what I have thought of home, and the dear friends I have left many miles from this. I can tell you that although deprived of every earthly connexion or even of a correspondence with them, and one hundred miles lies between me and the friends of my youth, still I am contented, still I am happy, the present witness of an indwelling God fills my soul, and I am walking hand in hand with a large circle of dear friends to Mount Zion the city of the living God.

My situation is as pleasant as I could expect. I have daily blessings heaped upon me. I am fed from day to day like the ravens,[13] and I can say to you to day I am happy in the enjoyment of the love of God and I anticipate one day though separated from the society of my friends here below, meeting them in the kingdom of God. Glory to God for religion that makes the soul happy, a religion that brings peace and tranquility will prepare the soul in the language of the Psalmist to say—"Though I walk through the valley of the shadow of death I will fear no evil—for thou art with me, thy rod and thy staff comfort me."[14] I left Lowell last May on account of my health and staid until Oct. in Boston and worked at my trade, except what time I was gone down on the water to Cape Cod. I went to Camp Meeting in August,

---

13. See Luke 12:24.   14. Ps. 23:4.

as usual was gone ten days, cast anchor three days—went ashore three miles from where we set sail, having in company upwards of two hundred, fourteen of which were Methodist Ministers. Had about twelve sermons preached on board, and one on the shore— dug clams—had plenty of good codfish, crackers and coffee—and on the eleventh day reached Boston wharf in better health and better spirits than when I left—having had but about six good hours sleep in ten nights. Just at this moment one of brother Rawson's Camp Meeting stories has popt into my head and methinks I hear him say, "Well Maria this is one of your Camp Meeting scrapes." Let me tell you my dear brother I love them now as well as I did five years ago. Yea far better—for I have known real good produced by them.

Time reproves me and I must draw to a close by saying dear mother do write to me immediately—dear brother and sister do write and let me know whether you are in the land of the living, whether you live in Killingly—whether you prosper in spiritual and in temporal things. As to myself I have enough of the good things of this life. I brought nothing into this world, and I expect to carry nothing out, a stranger and a pilgrim here.

My best wishes and most fervent prayers will ever attend my dear parent. Once more I say dear mother write to me, direct to Lowell, Massachusetts.

Your daughter,
MARIA CORNELL.

Letter No. 16

Lowell, July 4th, 1830.

To Mrs. Cornell.

MY DEAR MOTHER—I take this opportunity to acknowledge the reception of two letters, one of which I received last week. You say you should like to have me come to Killingly this summer. Last summer I made my calculations to visit you, and should have done so if you had written—but I have not thought very seriously of visiting you this summer, until I received your last letter. I then thought I should come immediately, but finding my engagements such that I could not be absent from here more than a week or ten days at most—I have concluded that the time I should stay would be so short—the expence would be more than it would gratify either of us. I am now preparing to go down on the water to camp meeting where I went last year. My health is tolerably good for the season, I never enjoyed my health better than after I went on the salt water, although I was very sea-sick. It is my intention now to spend two or three weeks with you in the spring, if life, health, and strength are spared me.

I have been in Lowell so long that I should feel lonesome any where else. My love to my sister, tell her I long to see her and the children. I shall write to Mr. Rawson as soon as I return from the Cape, though I never received a line from him or Lucretia since they were married, but I expect my sisters time is pretty much taken up with her children.

You will please inform me in your next if you have heard any thing from my brother James. The bell is ringing for meeting and I must close. I will send this piece of paper, it was thought it resembled me when it was taken—but I wear my hair in my neck short now, and it does not look so natural.

I am your affectionate though absent child.

MARIA CORNELL.

Letter No. 17

Taunton, March 10th, 1832.

To Mrs. Cornell.

MY DEAR MOTHER—I sent a little pamphlet to Mr. Rawson a few days since but I dont know as he will understand what I meant, I pitched my tent in Taunton last fall, about the time of the riot in Providence, I should have written before, but I knew I could not make you a visit in the winter, for this reason I kept still. I am now in very good business, and I do not want to lose my place, which I must do if I come to Killingly at present. You will probably wish to know what business I am in, I am hooking up, and folding cloth and keeping the weaving room books, I have the whole charge of the cloth and my employer is unwilling I should be absent even for one day, though I sometimes have two or three hours leisure in the course of the day, I think however I shall get leave to come and see you at Providence, if you could come there and meet me at uncle M—s, I will set a time, and I wish you to write me immediately— whether it will be convenient for you to come, I want to see Mr. Rawson and Lucretia, I hope I shall some time in the course of the present year. I will meet you in Providence the 18th of May, or 15th of June, just which will be the most convenient for you, I cannot leave the first or last of a month.

Your daughter in haste.
S. M. Cornell.

CHAPTER VIII

The circumstances detailed in the life of Avery, need little comment; every one must see in the persecution of Miss Winsor, and the slander of Rev. Mr. Norris, that Avery was a man of wicked, and revengeful, and persecuting temper; and his frequent closetings in the famous study with females, and the sad and grieved appearance of his wife, speak volumes. When apprehended for the slander against Mr. Norris, he was taken from the desk during a prayer meeting, by Mr. Kimball, a sheriff at Lowell, and was so much frightened as to faint, and several persons then made the remark, that he probably feared it was for some very different offence he was apprehended. But when put under arrest at Bristol for the alleged murder of Miss Cornell, it was said he exhibited great firmness, and during his trial discovered no signs of fear and but little agitation.

To pass any comments upon that trial after the able Strictures published by "Aristides,"[1] would look like vanity indeed, yet a few facts which have come to the knowledge of the author it may not be amiss to mention, particularly as many who read this may not have seen the ingenious and masterly criticisms of the trial referred to. Preliminary to the facts we are about to state, we will just make a short extract from that work.

"Never was a criminal trial instituted and carried through in this country in which so much baseness was manifest, so much chicanery practised, the public, the government, the court and the jury, so deeply insulted, nor an accused man acquitted with such a chain of circumstances against him. The whole machinery of the method-

---

1. Pseudonym of Jacob Frieze, a Boston clergyman, editor, and political pamphleteer.

ist church has been brought into operation and its artillery made to bear on the battlements of the hall of justice. Perjury, base and foul has been committed on the stand, under the sanction of a religious garb to protect a wretch from punishment."

How much of perjury was practised on the stand we are unable to say, but certain it is there was great exertion made to prevent witnesses testifying against the prisoner, by his friends the methodists; most unwarrantable means used to prevent the truth coming out. The circumstance related by Aristides respecting a sheriff of Newport having to run a race with a methodist minister, of nine miles, to see who would get there first, the sheriff to summon her or the preacher to prevent her, is strictly true, and that after all the vigilance of the sheriff the parson won the day and arrived there first, and when the sheriff came, the woman (a Mrs. Brownell if we recollect right) pretended to be too sick to go; what her testimony would have been if let alone we do not know, but if we are to judge of its importance by the violent efforts made to stop her going, we must presume it to have had great bearing upon the case. A similar instance occurred in Thompson, (Conn.) A Mrs. Patty Bacon, a witness for the prisoner, stated some circumstances which she said occurred at the Thompson Camp Meeting, of a very different complexion from the story told by her in Court. Mrs. Bacon's daughters, thinking it of some importance to the case communicated it immediately to the friends of the deceased, but before they or the counsel for the government could have a chance to converse with her, she had had a conversation with some of the methodist clergy, and when she was afterwards interrogated upon the subject, denied every word of it, and that she had ever said so, the testimony of her two daughters and son-in-law to the contrary notwithstanding, and was afterwards found on the stand testifying that "she had suspicions of the situation of S. M. Cornell at the Camp Meeting," to the amazement of her own family, who had never heard any suspicions mentioned from her before. The statement she made to her two daughters and son-in-law, was this: "That a very tall man with a dark frock coat, and broad brimmed hat whom she took to be a methodist minister, (she did not then know Avery) came to the Muddy Brook tent three times, Thursday, enquiring for Sarah Maria Cornell, and that she afterwards saw the same man conversing with her without the tent." All this she stoutly denied after the above-mentioned conference. This woman was a member of the methodist church. Again a Mr. Windsor, a respectable innkeeper in Dudley, was standing on the west side of the camp ground on the memorable Thursday afternoon near the time of the blowing of the horn, with a Mr. Jason or Judson Phipps, and a gentleman and lady passed them, when Windsor enquired who they were, and was answered by Phipps that it was a Mr. Avery and Miss Cornell, and added "I am watching them." Phipps afterwards in Windsor's bar room recalled the circumstance to mind, and in presence of several persons said, "that man was Avery and the woman Miss Cornell, I know them both." It got out of course that he had said so, and when the gentlemen in search of evidence for the government called on him, the following dialogue took place.

Question.—"Did you tell Mr. Windsor those persons walking together were Mr. Avery and Miss Cornell?"

Answer.—"I might and I might not."

Question.—"Did you or did you not say in answer to the question of Windsor, 'who are those?' say it is a Mr. Avery and Sarah Maria Cornell, and I am watching them."

Answer.—"I might and I might not."

The same answer was invariably returned, and it was all they could get out of him, until the gentlemen were obliged to give it up in despair. Nothing could be drawn from him.

Mr. Asa Upham, a sober industrious man, said to be a man of property and respectability, said he saw and knew Avery and S. M. Cornell, and saw them walking arm and arm together, in the woods near the camp ground. This person went to testify at the trial, and found the methodists had procured three persons to swear him down, and having no means there to testify to the character of these witnesses, he would not stay. Being an inhabitant of another State they could not detain him. What sort of persons the methodists had employed to testify against the veracity of this man may be gathered from the fact, that two of them were so intoxicated before they got half way from Providence to Thompson, as to be scarce able to continue their journey. We can state this fact without quoting "Aristides."

The attempts to brow beat witnesses in Court, to confuse and perplex them, so as if possible to cause them to falter or contradict themselves on the stand, was another most ungentlemanly, unmanly and unchristian proceeding, and was probably carried to a greater extent by the prisoner's counsel than has ever been attempted in any criminal case in this country, at least it is believed so by nineteen twentieths of the persons who attended the trial; and when the witnesses chanced to be persons of so much firmness that this was deemed impossible, either some one was brought up to impeach their character for truth, generally, or to swear that they had stated different to them at some other time.

Among the witnesses tampered with, there was none perhaps who underwent a more fiery trial than Mrs. Sarah Jones. As her name is mentioned several times in the trial, she will be readily recollected, but as the whole story cannot be perfectly understood from that that is told, and the whole is not related, we will give the narrative as she has given it to us, accompanied with her certificate to the truth of it.

The Rev. Mr. Drake, while enquiring in that neighborhood if any body had seen Avery, heard her say she saw a man in the morning; he wanted to know why it was not as easy for her to say it was in the afternoon as morning. By this means Avery and his counsel became possessed of the fact that Mrs. Jones had seen a stranger pass their house in the neighborhood of the coal mines,* or rather on the route to it, in a very early state of the business, directly after he was put under arrest, and previous to the examination at Bristol, and they sent for her to come to his house. When arrived there, she was very cordially greeted by Mr. Avery, who introduced her to the presiding Elder as "one of his witnesses, who *saw him on the Island,*" and she was asked to relate the circumstances in presence of his attorney and the others. This she did. She had been looking out all the morning for the return of a brother who had been living at the eastward, and was expected back on that day. Between 11 and 12 o'clock she saw a man come through the white gate, and come within ten or fifteen rods of her. She described his course by answers to questions, gave an account of his route, and of the country, which the counsel traced by chalk marks on the floor as fast as

---

* Perhaps there is not a set of people in New-England more primitive in their manners than some on this part of the island. The following lines were composed it is said by an old lady over eighty years of age, in the neighborhood of the coal mines. We do not know when we have taken up any thing that sounded so much like olden time; if it amuses the readers of this book as much as it did the author, it will well repay them the trouble of reading it.

"Young virgins all a warning take
   Remember Avery's knot [spelt not.]
Enough to make your heart to ache,
   Don't let it be forgot.

You mothers that have infants
   To sympathize and mourn,
Such murder never was done here
   Or ever yet was known.

He killed the mother and the child,
   What a wicked man was he;
The devil helped him all the while,
   How wicked he must be.

He dragged her around upon the ground
   Till she no noise could make
Contrived a lot—tied Avery's knot
   And hung her to a stake.

The devil he was standing by
   A laughing in his sleeve,
It is so plain he can't deny
   He must not be reprieved.

He preached the gospel night and day;
   What a wicked man was he;
The devil helped him to preach and pray;
   How wicked he must be.

How could he stand to preach and pray
   With murder in his heart;
The devil helped him day by day,
   And he will make him smart.

Methodism he did profess
   For that was his belief,
How can he ever take his rest,
   He must not be reprieved.

Hang him, hang him on a tree
   Tie around him Avery's knot
Forever let him hanged be
   And never be forgot."

she described it. Thus did the simplicity of this woman furnish a pretty correct map of the country. But unfortunately, the man she persisted she saw in the *forenoon;* and when she came out, Avery, who went with her to the door, said in passing the entry, laying his hand on Mrs. Jones' shoulder, and looking imploringly in her face, "My life is worth more to me than a thousand worlds, and my life depends upon my witnesses—can't you recollect Mrs. Jones that it was in the afternoon?" but, "say nothing," he added, to which Mrs. Jones answered she would not, and as she says kept her word until circumstances made it imperiously necessary she should disclose this interview, and the conversation that took place.

On the day that Mrs. Jones' testimony was required, at the examination of the prisoner in Bristol, she was brought over and was stopped at the house of a Mr. Tilley in Bristol; where a Reverend gentleman met her, she says, at the door, exclaiming, "now Mrs Jones, *you must remember it was in the afternoon* when you saw the man, for Oliver Brownell has just sworn he saw the same man, and it was in the *afternoon.*" For a moment she said she felt almost bewildered, but the firm conviction that she had stated nothing but the truth, and that if there were ever so many men of that description seen in the afternoon, the one she saw was in the morning. Directly some one came up to her and shook hands, saying, "we have been to tea, and Mrs. Jones here has not: you will be so good as to get her some, will you not—as soon as possible?" and the good sisters hurried to get her tea, overwhelming her with civilities. The tea was already on the table, and the lady about to partake of their hospitality, when she was called for to the court. Two of the daughters of mine host volunteered to accompany her, thinking she would feel intimidated to go without any female, and on the way, short as it was, endeavoured to influence her to say it was in the afternoon. One says, "well you saw brother Avery, it seems, on the island?" "No, I don't suppose it could have been him," said Mrs. Jones, "the man I saw was in the morning." "Oh, it must have been him," said one, "it could not have been any body else; and you must try to remember it was in the afternoon." By this time the trio had arrived at the scene of action, where the matter was put upon oath; and Mrs. Jones described the stranger, who really, from her description, must have borne some faint resemblance in person to the prisoner. But alas! the stubborn witness would not say it was in the afternoon: after all the examining, cross examining, and twisting of evidence, nothing could be got out of her but the same old story, "it was in the forenoon." Was there ever such obstinacy heard of? that so many civilities should have been thrown away! But so it was: and the woman was conducted back to the house of Mr. Tilley, where she had engaged to return to tea, hurt and abashed by the altered looks of her now silent guides. Nothing was said except as one looked mournfully upon her, and, as she thought, reproachfully, and said "we were in hopes you would have *remembered it was in the afternoon.*" Poor Mrs. Jones went in with a heavy heart, feeling that she had disappointed the hopes of the prisoner's friends, but (unable by any sophistry she could imagine, to make out that between eleven and twelve o'clock was in the afternoon) with an approving conscience. She said the young ladies passed into the other room and were followed by one and another. There was a whispering conversation going on there, and each, upon returning, would eye her with scornful and angry looks. It seemed, she said, as though the tea never would be ready, but at length she was called. "Never," said the poor woman, "did I eat a meal before that I thought was begrudged to me." But at length, she said, she took courage, and feeling tired and faint, resolved to "drink as much tea, and eat as much as she wanted to:" directly after which she took leave of her now ungracious hosts and went to a tavern, and staid all night, and rose early on the following morning and returned home:— a distance of four miles including the ferry, which she had to give eight cents to cross. From this time until after Avery's flight, and his being taken again, nothing could exceed the scornful and supercilious manners of the Methodists to this woman, by her description, whenever she met any of them. Her own expression was, "they turned up their noses at her, and would not speak." But when Avery, after his flight, was pursued and taken again, to her amazement, all at once their manners changed. Whether they had pouted it out till their resentment had worked itself off, or whatever was the cause, they now began to relax the muscles of their faces, and not only to give her the time of day, but even to shake hands very cordially and enquire after her health. Behold a polite letter arrived from Bristol, dated 28th March, enclosing three dollars; the letter states that it encloses the fee for travel and attendance in the case of the State against E. K. Avery, but does not mention the sum—the sum enclosed was three dollars, she states, and it appears she opened it in the presence of another person. When the trial came on at Newport, she was again summoned by the prisoner to testify. She was in the State and was obliged to go. The person, a methodist, who went to carry her, and she said to him, "what did you come for me for my evidence can do Mr. Avery no good, for the man I saw was in the morning." "Why we were in hopes Mrs. Jones, you would *remember it was in the afternoon*," was the answer. Arrived at Newport she was conveyed to the house where the Lowell witnesses were quartered, where she was again hampered to say it was in the afternoon when she saw the man on the island. She said the witnesses were shut up together in the front room of the house, and practising most of the evening to try to make the clove hitch,[2] the Whitney girl and all, and that they asked her to show them. "I cannot for I never saw one made in my life," she answered. "I did not state this circumstance to the court," she said in her narration, "because I did not then think it of any importance, but when I found one of those very girls came forward in court and swore she had been used to seeing it in making harnesses, and showed how it

---

2. A sailor's knot, formed of two loops facing oppositely, in which the rope ends are passed around and through the loops, emerging in opposite directions.

might have been used by the deceased to hang herself, I then regretted extremely I had not told that this very girl had been drilling to practice that manœuvre all the evening, and that they did not, when I was with them, appear to understand how to make it, and asked me to shew them."

Unable to twist the evidence of this woman to suit their purpose, the friends of the prisoner endeavoured to make it appear, on the stand, that she had contradicted herself once or twice in conversation: but they did not make it out very clearly; although it was a subject of amazement to many that she had not done it repeatedly, placed, as she had been, in circumstances of such embarrassment and temptation. Since writing the narrative the author has been warned by some of Avery's friends not to place any reliance upon any thing this woman should tell, as there would not be a word of truth in any of it. But when we wished to see the letter from Bristol, in confirmation of that part of her story, and she produced it; and after ascertaining from the people of the house, that she was not only on that evening, at the house, with the Lowell witnesses and the girls named; but that they were *shut up by themselves in the front room,* the very words she used; and that those girls were repeatedly seen practising upon that knot while there; we could not but believe it: particularly, as we have never in all our travels been able to find any one who used the clove hitch in harnesses, and have seen at least hundreds making.

Mr. John N. Smith, who testified to the cord being different from that used in factories, was urged to go as a witness for the prisoner, which he refused because he knew his evidence must be against him, of course. One person, a Methodist, and if we recollect right a deacon in the meeting, urged him to go to Newport to testify for the prisoner. Said he, "you will be at no expense—and here is a five dollar bill, if you will go." We asked leave of Mr. Smith to state this fact, saying such things ought to be exposed. He objected, saying he was ashamed to have it known that any man should dare to offer him a bribe.

It must be evident to every candid observer, that the testimony for the prisoner in many instances was overdone. For instance, had two or three respectable persons of good standing in society stated that the character of S. M. Cornell was not good, and that she was plotting, revengeful, &c. it would have gone farther toward convincing the minds of the public than all this array of questionable evidence; a great deal of it was entirely irrelevant to the case; a vast deal appeared to have no object but to blacken the character of one as we observed before, who was *where she could not answer them back again,* and injured in a very material manner the credit of a society who could tolerate such a character (allowing that she was so,) so much as to retain her among them, to be on any terms at all with her. To receive *again a woman upon probation,* who had once been expelled upon *such a charge* as Doctor Graves made against her. Gracious heaven! the idea is monstrous—the thing incredible, if they had not stated it themselves about themselves. Who that reads and believes such a statement can be willing their young and innocent daughters should be followers of a meeting proverbial for their familiar and social habits, where they would be liable to associate with such characters.

We do not know but a part of the charges against her may be true, because we have no means of positively knowing; some of them we know cannot be; for instance, we know she could not have been sinning and playing the hypocrite in four different places at once, as is found to have been stated in one part of the trial. She could not have been at Dover, Great Falls, Lowell and Waltham, at one and the same time, without possessing one of the attributes of omnipotence—that of ubiquity.

We scarcely think she could have been the writer of these letters, which we know she was, if as vile as represented. Oh exclaims the scoffer who reads them, and believes the account given of her at the trial, what a caricature they are upon talking, canting, whining christians. But to those who hope better things, what a different aspect will they wear. To those who believe that "out of the abundance of the heart the mouth speaketh,"[3] they will not appear like hypocrisy. Ignorant and enthusiastic we allow her to have been. Ignorant as she was though, we observe her letters are much better spelt than those of Avery which we have seen.

We are sorry to say that what we have stated respecting the treatment of the witnesses, together with much more, generally known, which our limits will not permit us to state, goes far towards contradicting the assertion made in the report of the Conference, who sat upon Avery's last examination, viz. that while the trial was pending they remained perfectly quiet, not even undertaking to clear their brother from any of the ridiculous and exaggerated reports daily circulated against him, or to contradicting their reports. It is apparent it was no time to stand up in his defence in that way—but as to remaining quiet, it will be seen by every body that they were as busy as *moles*, all the time. It is amazing that Avery should not have had the politeness to publish a card afterwards thanking his reverend brethren generally, for it would have been hard to have particularized names, where so many deserved the meed of thanks, for their great exertions and important services!!! Have the proceedings of these reverend gentlemen resembled those of a religious association? Has it not rather looked like a combination of men for secular and political purpose—a league offensive and defensive? Has it appeared their object is to elicit truth, or suppress it? How is a charge against one of them treated? Is there a candid examination of facts gone into, or is not every movement directed to break down the character of the accuser in the first place, or to invalidate his testimony some way or other? Is this the way to come to truth?

We come now to the last remark except one we have to make on this painful subject, viz. the subject of those letters found in the possession of S. M. Cornell, designated as the yellow letter, the pink let-

3. Matt. 12:34.

ter, and the white one. (The original letters now in the custody of the court, have been kindly and politely submitted to our inspection.) The view of the author in seeking to see them was simply to ascertain for a certainty, whether the deceased had any hand in them, as insisted upon by the friends of E. K. Avery. To some of them, honest, though prejudiced people, we had pledged ourselves if possible to obtain a sight of them. Having been employed for many days in transcribing her letters into this work, we felt perfectly confident, that if she had any hand in them we should at once detect it, however disguised. Among the papers found in her possession we discovered nothing in her hand, however, except her letter to Mr. Bidwell, and the slip of paper containing those words—"If I am missing," &c. The three other letters were written by one person, although the pink one was written very fine, and disguised to make it resemble a woman's. But oh when they were compared with the acknowledged letters of Avery to Mr. Bidwell, Mr. Drake,* and Mr. Storrs, the conviction which they brought to my mind was absolutely overwhelming. We thought we had fully believed in Avery's guilt before, but we feel we never had, until then, a gush of feeling which we could not prevent, choking utterance for some moments. We do not wonder that his friend Mr. Bidwell could not help saying that one of them evidently was his hand writing. That one of them was the plainest, but they all discovered one common hand, all the peculiarities, the turn of the letter, the dash stops, the breaking of some and the leaning of others, the spelling—of folding and sealing, even to the most minute particulars, was exact in accordance, one thing too as judge——, observed—was very convincing. "If any one

---

* The methodist minister preaching in Portsmouth. There is something worthy of notice about this letter. It is dated the 22d, and consequently was written on Saturday after the murder, and before Avery knew that he was suspected of it, and most urgently requests Mr. Drake to come to him immediately, and without delay; to come horse back or in carriage, or any way——and all expenses should be paid, but to come without fail. This Mr. Drake was remarkably busy during the whole trial. Nothing could exceed his zeal in serving the cause.

had forged these letters intending to have them attributed to him, would they not have put his name, or, at least the initials of it?" but instead of that they were signed "B. H." for Betsey Hill, probably, though it seemed the writer lost his recollection in one place, where he says, "direct your letters to Betsey Hills, and not to me."

The fact is it was confidently anticipated by the author of these letters, they would never come to light, she had been directed so positively to burn them. One said, however, "you may keep the letter till I come and bring them and I will bring mine." It was thought no doubt she had them with her in the pocketbook or wallet which she always carried in her pocket, and in which she had generally carried these letters. It was stated to the author that when found the pocketbook was not about her, and that from that day to this it has never been discovered. It is amazing this circumstance has never been commented on before, for if true undoubtedly the author of those letters, whoever he was, took it from her, expecting it contained the letters, which it seems she had taken out and put in her bandbox. In her pocket was found a silver pencil case, and some other trifles—we have forgotten what—but no pocketbook. It would appear singular, if we did not recollect that the finding of the vial of tansy oil was not testified to in Court, or much said about it until within a few months past, although the women who laid her out and found it among her things, talked about it at the time. Mrs. Nancy Durfee testified to the author that she was the first person who saw it in the trunk, where it lay with a teaspoon beside it, and that from the quantity she could not believe it to be oil of tansy— but such it afterwards proved. We believe there is one person who knows who sold this vial. We have ascertained she did not procure it herself at Fall River, nor carry it there with her. There is a peculiarity about the vial, which we believe would cause any person who sold it to recognise it again.

It is almost equally strange so little should have been said respecting the wounded hand of the prisoner. At the time of the murder a woman in the vicinity of the place, dreamed that the murdered girl appeared to her and told her that the person who killed her, might be recognised by the marks of her teeth upon his hand, for that during the struggle he put his hand over her mouth to stifle her shrieks, and she bit it. This woman determined to see the prisoner, and know whether there were any such marks on his hand or not. Whether it was owing to her entreaties or not we cannot say, but at the Bristol examination the prisoner was ordered to unglove. He had kept one glove on, previous to this. He pulled off his glove and his hand was found to be wounded. The counsel for the government wanted a physician called to examine it, which the justices who sat on the examination declined to do, when Avery offered to account for the wound by relating the manner in which it was done. He was silenced by the government counsel, who did not wish to hear *his story.* And thus the affair of the wounded hand was dropped.

In conclusion, we would observe, that however strong the presumption of the guilt of E. K. Avery may be on the public mind, we fervently hope he may remain unmolested, and we would wish unnoticed. If he is guilty, the avenger of blood is behind him. That is sufficient. If innocent, he ought not to preach. Still, silent contempt and utter neglect would do more towards putting down such persons than clamour. Mobbings never ought to be known in a civilized community; besides that they are calculated to make even merited chastisement appear in the eyes of the world like persecution. They have another horrible tendency, which is to give men in power who are fond of the exercise of authority a pretext for the most shocking severities. Those who set out to inflict chastisement in this way generally get the worst of it before it is over. Besides, who knows that in this case that is not the very point aimed at? Who knows that the outrageous insult of thrusting this man into the pulpit in every place where he is most obnoxious is not intended to produce this very result? Should a meeting-house be pulled down, or preacher torn from the pulpit, who knows what fearful war cry might be raised. There is nothing whets the sword like false zeal. The battles of the Cavaliers and Round Heads[4] might be fought over again in our country, experience has once demonstrated there is no quarter to be expected from men who would march to battle singing psalms, and wield the battle axe with the covenant in their bosoms. From all such contests may the *God of peace* deliver us.

To return to Fall River, that place from which sin and sorrow and contention have kept us away so long in story. Though its natural beauties, as we observed before, are obscured by improvements, it is, and ever will be, beautiful in situation. The waters of Mount Hope bay still roll on in their natural course.

> The waves still wash the peaceful shores around
> Where the poor wanderer a grave has found.

Internal improvement is going on, and wealth flowing in. But there is a change there—and oh, how great a one! There are two classes of people who once lived in friendly intercourse, between whom, now, nothing but frozen civilities are exchanged. The great body of the inhabitants must feel, as every one of our republican states would have felt, if opposition had been enabled to palm upon us the curse of a people, within our own borders, having a seperate and independent government within themselves. Can nothing be done to heal the breach? Children of the same heavenly Father, redeemed by the same power, can no compromise be effected? Alas! we fear not. Fanaticism, aided by self-will and obstinacy, has drawn the sword and thrown away the scabbard. The minority have nailed the flag to the mast, and are determined to surrender only with life. Ah! foolish and perverse generation; something may yet happen, to convince you of your error: but your conviction may come a day too late. How mournful it is to see people, who once exchanged a friendly greeting whenever they met, even in the streets, with a cor-

---

4. Supporters of King Charles I and of Parliament, opponents in the English civil wars (1642–49).

dial "good morning," or a friendly shake of the hand, now pass each other without any salutation, and perhaps even with looks of coldness and contempt, of estrangement and aversion.

Yet the hospitable and benevolent inhabitants of Fall River, as a body, are not in fault in this case; they have only sought to do their duty towards a helpless stranger who perished by lawless violence within their precincts. And for this, their reward shall be the eternal hatred of one particular class of people, and the esteem of every honest, candid and impartial person.

If Fall River was once an object of interest to the traveller, it is doubly so now, from the associations connected with it. For months after the tragical affair detailed in the preceding pages, it formed the entire topic of conversation in every steam boat that plied the river; though time has effected a change in this respect, yet ever as the boat nears the bay is the stranger heard to enquire the situation of Gifford's house, of the road to the ferry, of Howland's bridge, and of Durfee's farm. It is in vain for the peace professing part of the community to say "it is time this subject was dropped—this excitement ceased." To get out of the way of it we must go as a certain writer says, "where a *hay stack* was never heard of."

APPENDIX

So much has been said of late of Camp Meetings, and such intense curiosity excited on the subject, that the author of these sheets feels called upon to give a history of one of which she was an eye and ear witness, i. e. for the time she passed there. The meeting was held in R. I. and was I should say some ten or twelve years ago. It is said that the regulations of those places have been much more strict of late years, and that the disorders in the immediate vicinity of a Camp have lessened since. That their moral tendency is better than it was before is however doubtful. Witness the affair of the unfortunate girl who perished at the stackyard.—That her latest misfortune was occasioned by her attendance there, cannot be doubted. The testimony of her sister and one other respectable female fully proves that; even by the testimony of her enemies it appears she had made resolutions of amendment, which the temptations and facilities of a Camp Meeting overcame. In fact it is asserted by many that no certain proof exists that her first criminal offence was not perpetrated there: that might have been the beginning—the end no one disputes was death. Nothing can be more imposing than the first view of a camp, and a superficial observer, a person who entered and just walked through, or was so fortunate as to be seated in some safe place while listening to a sermon or a prayer, might see no harm in one. But we think no person could pass much time in one, if a person of any observation and not blinded by fanaticism, without deprecating the practice. I am fully aware I shall make no friends by an exposure of all I saw and heard there, but I hope no enemies. I hope, fervently hope, that no order of men have become so depraved, as to hate or persecute any one who dares to avow a difference of opinion, or for speaking the truth, however repugnant to their own views or feelings. Should it be asked, where has this history been all this time? I answer, safe locked up in my desk. Why has it never been published before? because it has never been called for: the occasion which has called it forth has never been so pressing. Men's eyes are now partially open to the great evils of fanaticism generally, and of Camp-Meetings in particular; and every thing known on that subject ought to come out. The following diary or memorandum, or whatever it may be called, was taken at the time, except a very little added from memory. Were it necessary, I could give the names of many of the characters mentioned: but it is not with a view of injuring individuals, or dragging before the public names of persons known only for their modesty and domestic virtues, that this is brought before the world; but with the hope that it may have a tendency to assist in putting down a great evil, a sore affliction in the land, a pestilence walking in darkness, an enormity that calls loudly for the strong arm of the law, in the opinion of many good judges. A thing much more to be dreaded than even theatrical entertainments, inasmuch as it goes under the name of religion; whereas the former is called by all sorts of evil names that can serve to warn people. When people go to the theatre they know where they are going. They go with their eyes open. They know it is at best but a profane entertainment, and they go against the warnings of the pious of all denominations. But when they go to a Camp ground, they do not know of the dangers that lurk there and menace them at every step: they do not know who or what mingles with the motley assembly that surrounds them. They are told that by going there they may *find religion*, (a most absurd phrase by the way) as though religion could only be imbibed in certain places and situations. They are taught that the spirit of God, whose still small whispers[1] may be heard at all times when the soul is disposed to

---

1. See 1 Kings 19:12.

retire within itself and listen to its heavenly breathings, that the spirit of God is in a very peculiar manner dispensed at those scenes of noise and confusion, even without measure, if we may credit their often uttered expressions of "I'm full," "I'm running over," &c. Many really honest people are induced by the hope of being converted there, without any trouble of their own, to attend frequently, to the great detriment of their families who too often are suffering for their care at home.

The feelings of the really pure and pious and intellectual among these assemblies must often be outraged—though they endure it thinking they are doing God service, and it must be right because their councils permit or decree it.

Mistaken beings, there is a way that seemeth right to a man, but the end thereof is death.[2] I am fully aware of all the arguments made use of to defend Camp-Meetings—but they are futile; and fully aware as will be seen by this of their imposing effect at first sight.

Who can contend that this free intermingling of society is not dangerous, this tumbling and falling about not indecent. That the familiar habit of life practised there is not full of temptation; to prove this would be to prove that the persons who frequent them are not made of flesh and blood, a thing that abundantly proves itself.

Who can prevent the neighborhood of a Camp Meeting from swarming with drunkards and gamblers, and horse jockies and pick-pockets, and offenders of every other description, who go about seeking whom they may devour. It is said now, that "bad people are driven off the ground," but if so where do they go to? it is certain that hack loads and wagon loads of very bad people are always seen following a Camp Meeting as regular as to a field of battle. If it is purposely to call sinners to repentance that these out door meetings are held, why not have a place assigned them, where they may hear and be profited by the preaching of the word—and kept in sight

---

2. Prov. 14:12, 16:25.

that people may know what they are about, rather than be driven into the bushes to pollute the place with all sorts of enormities. I am not however contending for their admission at all, if their presence could be avoided, for in my opinion the best place to preach to these people is at home, and admire the plan of domestic missionaries to seek out the abodes of vice. But if the vilest sinners are not to be benefitted by these meetings why not hold them in places of worship where no facilities for crime exist?

But leaving the out-door evils of Camp Meetings entirely out of the question, it still remains to be asked, are their in-door evils not to be dreaded? is their effect upon the religious society who frequent them less pernicious? are not habits of idleness and dissipation (spiritual dissipation though it be) promoted by it? among fanatics especially—among those whose business is at home—among those whose feet ought to abide in their own house? If we were like the wandering Arabs or Tartars whose home is in the fields, in every green spot whereon they may chance to light, it would be a different affair, but in a country like ours where domestic industry and sobriety are of such importance, wandering, idle habits among females are absolutely ruinous. Add to this the exposure of health. There can be no doubt that the lives of many delicate females have been sacrificed to the absurd custom of sleeping on the ground, with no covering but a tent to those who have been accustomed to the walls of a house. The want of rest, of sleep, which all who attend these meetings must suffer more or less, must be great. And last but not least, the low and dishonoring thoughts of religion which the constant hearing of such familiarity with Deity must unavoidably create. Why it is impossible people can know what they are saying, when they use such expressions as are frequently used at these places. They frequently speak of the Almighty, and speak to him too, as though he were an equal, and even an inferior; for people very seldom address an equal in the imperative.

We have one thing to hope, however, if people will not submit to be reasoned with; that as the light of science breaks upon their minds, bigotry, superstition and fanaticism will vanish. Unless, indeed, it be a combination for civil purposes, as in the case of the Romish Church, where, as her clergy became more enlightened, they only became more accomplished to do evil; and in proportion to their knowledge, so did their tyranny, and extortion, and oppression of their simple hearers, increase.

It is possible that some very hard thoughts and still harder speeches may be the reward of the writer from those who differ in opinion, but why? How many have we heard of the Methodist denomination inveighing against Masonry—not because Masonry ever did them any injury, but merely because it had become so obnoxious to so large a proportion of the people—and contending that Masons ought to give up their charters, if it was only in compliance with public prejudice and to restore general harmony. This was certainly sound and correct reasoning: but it applies equally well to Camp Meetings, and to the case of E. K. Avery.

Extracted from a Journal of a Camp Meeting, held in Smithfield, R. I.

The long expected time at length arrived, the meeting was to be held in an extensive wood about nine miles from the town of —— several very respectable young ladies had agreed to stay at a house within two miles of the meeting, where they could ride backwards and forwards as often as they chose through its continuance. The weather was excessively warm and the season unusually dry, and from dust and heat, most uncomfortable riding. For my own part I felt determined to endure all hardships rather than be disappointed in this opportunity of seeing and hearing. So many stories had been told me of Camp Meetings, and such various and contradictory ones, that I felt determined to see and hear for myself. The meetings had not commenced upon our arrival, but the Camp was said to be in order, and so great was our impatience to see it, that we accepted of a ride with a company of friends who were going from the house where we staid, and proceeded to the camp ground.

There was an avenue opened through the wood from the road to the camp, perhaps of about a quarter of a mile in length, but as it had just been cleared, was exceedingly dangerous riding. However, we arrived safe at the entrance, and dismounting, passed the barrier, and found ourselves within the circle of the camp. I was never more amazed than by the scene before me. It was a beautiful spot in a pine wood. The trees were felled here and there with a sufficiency left for shade, and had the appearance of a fine grove within an impenetrable wood. The spot prepared was entirely round in shape, and its circumference I do not know, but it was quite extensive. The setting sun lent its last bright beams to the scene, while the snowy tents stretched far and wide, discovered many happy faces peeping from beneath their white curtains. Here and there an old man or woman was setting in the door enjoying the refreshing odour from the pines, or rapt in contemplation of the scene before them, upon which they appeared to gaze with much pleasure.

I was lost in admiration: a holy calm took possession of my soul: I thought of the camp of Israel—of Abraham sitting in his tent door in the cool of the day—of the patriarchs of old, who, as the inspired historian informs us, were "plain men dwelling in tents."[3] My imagination ran through the whole scene of sacred history, from Adam down to Moses—the plains of Mamre—the desert of Sinai[4] was before me—I heard in imagination "the trump of God, and witnessed the proclamation of the law."[5] In short it was some time before I could descend to earth and seriously consider the object that brought me there.

---

3. Gen. 25:27.   4. *plains of Mamre:* Gen. 13:18; *desert of Sinai:* Exod. 19:2.
5. See Exod. 19:16–20:18.

The plain dress of the people was very pleasant to me, and about the place there was an air of quiet, inviting to heavenly contemplation. And is this, I asked, a Camp Meeting? I do not believe a word about the confusion. Ill natured world that it is, what can be more proper than to retire into the desert to pray? What more likely to keep in mind that we are strangers and pilgrims here, than to become the inhabitant of a tent for a certain season? They came doubtless to recollect that "here we have no continuing city"—to remember that we "have no certain habitation but that house eternal in the heavens, whose builder and maker is God."[6] And here too, doubtless, they came to mourn for the sins that made us pilgrims and exiles in a world that but for our disobedience would still have been man's Paradise. In this state of exile, of humiliation, they "eat their bread with tears, and mingle their drink with weeping."[7] What can be more useful? what more proper? what more salutary? Surely my God must look with complacency upon a scene like this.

With a heart too full for conversation, I walked around the ground. My companions too were not very loquacious. I imagine the scene struck them something as it did me, except one who was not so ignorant as myself. To her I made the remark, "it is very quiet here." She answered, "the meetings have not began."

Upon re-entering the avenue our wagon had to turn out often for companies of rude young men, who, though the pass was so extremely dangerous, drove Jehu-like,[8] unmindful of stumps or stones, and appeared in a high frolic. I inquired the meaning of this. The landlord's son who was driving us, answered, "They were professional gamblers and horse jockies, who followed a Camp Meeting as regularly as crows and vultures followed an army." I was amazed, but I soon forgot the circumstance and relapsed into my former pleasing reverie.

---

6. Paraphrase of Heb. 13:14, 1 Cor. 4:11, and 2 Cor. 5:1.

7. See Pss. 80:5, 102:9.   8. See 2 Kings 9, especially verse 20.

Tuesday, we rode into the camp ground immediately after breakfast. Upon arriving in sight of the tents, I remarked to one of my companions, we had never seen an army encamped, and I hoped we never might, but this must resemble it in appearance except in warlike preparations. But here, exclaimed I with enthusiasm, they wage a holy war. These are engaged in a warfare that must never end, and their vigilance must exceed that of those whose sentence is death for slumbering at their post. Their foes are from within as well as from without; but they fight under a powerful leader, and be their foes ever so numerous, the banner of the cross will finally prevail.

Under the impression of such feelings, I entered the circle ready to join the people in their holy work. The ground is what is called an inclined plane, that is of rather gradual descent, and towards the lower part was erected a small platform, intended for the pulpit, called familiarly "the preacher's stand." A man was standing there when we entered and loudly calling upon the people to repent: there was little sense or connection in what he said, but he seemed to be very earnest and sincere, and had I think, the loudest voice I ever remember to have heard, but his exertions seemed thrown away, for except a strolling party occasionally halted near him, he had no auditors. We agreed to take a walk round and see what had become of the *good people*. They were mostly in their tents, cooking and eating, and in another apartment (for they were usually divided into two or three) reclining on the straw—men and women promiscuously chatting and laughing, and sometimes casting a furtive glance towards the preacher, whose extreme earnestness apparently excited little interest. After a time he was succeeded by another, equally vehement; and one old man began to cry aloud for mercy, which seemed to encourage them very much. Being fatigued, we retired to one of the tents where we had some acquaintance, to rest. They informed us that none of their best speakers had appeared yet, and that the evening was the time for powerful meetings. Towards noon, we observed people gathering round one of the tents, and following the multitude we entered.

Curiosity had been excited by the falling of a young woman on the ground near the stand. She had been conveyed into this tent, and was now lying on the straw, while the people who brought her returned to the stand, and seemed to take no further thought of her. We approached the young woman and felt her pulse, and believing she was in a hysteric fit, and that it would be highly injurious to let her remain so any time, begged permission to employ restoratives; but to all applications for them, and remonstrances, they turned a deaf ear. The people within smiled scornfully at my ignorance—told me she was happy and it would be a sin to revive her if they could. She was very pale and her pulse very low; but upon my persisting in rubbing her and calling for restoratives, backed as I was by a skillful physician who now entered the tent, and was remonstrating in not very gentle terms upon trifling thus with human life, they ordered us out of the tent, and fastened down the curtain, which excluded every breath of air, in an intense hot day, with orders for no one to disturb her until the Lord chose to dissolve her trance.

From this spectacle we retreated towards the African tent, which was filled with coloured people, but there was so much talking there at once, and they were so thick, we were obliged to pass on. A very old man now took his place on the stand, whose hoary head and stooping gait proclaimed that time with him would soon cease to be. I observed to the young lady who had my arm, that "this old man who stands upon the borders of eternity, will certainly feel what he says," and we descended again to the stand.

He commenced his discourse by saying—"he was a very old fashioned Methodist, and he should not be put out if they should *groan*, or say *amen* or *hallelujah*—that he had seen nothing that looked like zeal among them yet—no efforts to take the kingdom by force," &c. &c.—and he exhorted them with a degree of violence, which soon exhausted him and compelled him to yield the pulpit, i. e. the stand, to another. It had seemed the preacher's object in this discourse to bring his audience to a certain temperature, and I was lost in reflection, thinking if the man could mean that the Lord was not worshipped with acceptance where he was worshipped in silence. But I had little time for reflection on the subject, for a very devout looking person now advanced and requested his brethren to join him in prayer for the multitude. Being hemmed in in such a manner it was impossible for many of us to kneel, and many doubtless were afraid to. It is true if we had been engaged just as we ought to have been, we should not have seen what we did see. But this was impossible without absolute danger to ourselves. We had just been warned not only against pickpockets, but told women were often grossly insulted there, even in the thickest of the camp. Our eyes were therefore about us, and several young ladies afterwards told us that whenever they closed their eyes, and tried to engage in prayer, they were aroused by some of the men pressing so near, they could almost feel the pulsation of their hearts, and sometimes press their arms, &c.

But our greatest astonishment was to see the Methodists themselves wandering about in all directions, and some that were kneeling near us did not in the least appear to be engaged in what they were professedly about. One young woman, who knelt by our side, was busily employed in trying to fit a piece of bark to a log, with a countenance that expressed anything but devotion. One of our companions, who was watching her, burst into a violent fit of laughter, which she seemed unable to restrain, although we gave her very severe looks and shook our heads at her. She was not a professor of religion, though a most amiable person and a sincere well wisher to the cause; and she appeared to be very much mortified upon being told of it afterwards, though she assured us, with tears in her eyes, that if her existence had depended upon her suppressing it, she could not have done so.

The excessive heat and fatigue drove us back to our lodgings at noon; but towards night we rode again to the Camp. We observed, as we came near the wood, the recent erection of stalls to sell liquors and refreshments; and around many were congregated people notorious for dissolute morals and disgraceful conduct. The wood appeared to be swarming with people of all descriptions, and it looked as though it might be extremely hazardous for any one to venture there alone and on foot.

The first object that met our eyes upon coming within the barrier was a young woman of extreme beauty, who was staggering through the Camp, with her clothes torn and her locks dishevelled, wringing her hands and mourning that the people were not more engaged. She was a girl of about middling height, rather fat, with large, languishing black eyes, and a profusion of raven hair which floated on her shoulders and reached below her waist, with the fairest complexion that could be imagined. She appeared to excite great attention wherever she moved through the crowd. We observed, as she passed along, that the young men exchanged winks and jogged each others elbows. We subsequently saw the same young woman lying in a tent, apparently insensible, i. e. in a perfect state of happiness, as they assured us. There was a great deal of joggling, pinching and looking under bonnets, which was extremely annoying. We met a young lady from our town, who showed us her arms pinched black and blue by she could not tell who, while she was listening to the preaching of a woman at the stand. She was quite enraged about it, and protested she would get home as soon as she could get her party to go, and that no persuasion should induce her to come again. Her arm really appeared in a swollen, bad state. She was a woman of very correct deportment, and the conviction that no impropriety on her part could have been the provocation to insult, rendered the circumstance rather alarming; and we resolved to keep very near our friends and return home at an early hour: though various persons of the meeting tried to prevail on us to stay, saying "the work of the spirit was much more powerful after dark." There appeared to be a great deal of uncivil amusement going on, not only in a sly way, in the Camp, but throughout the ground. The narrow, dark avenue was exceedingly hard to pass; dissolute and drunken people were frequently in the rear of the carriages, swearing and talking in the most profane and indecent manner. Upon retiring for the night, I had a most serious time of self-examination whether it was right to go again; but the desire to know the extent of the evil or good prevailed, and I resolved to see it out, as the phrase is. Wednesday, however, I did not go at all, being confined with a violent headache. There was no rest in the inn; constant quarrelling in the road; the men very profane, talking every thing. The Land-lord of the house where we were, a very quiet man, appeared exceedingly annoyed.

In the course of the day, we heard often from the Camp. It waxed warmer there. Many were struck down, they said, with conviction of their sins, throwing themselves in the dirt and calling loudly for mercy; and many more *lost their strength:* the state of exhaustion described in the preceeding pages. The people without, we were informed, became more noisy and obstreperous. "Oceans of rum," as it was often expressed, were drank in the neighborhood. All that night we slept but little. Some of the profane lodgers, on the other side of the building, were continually singing hallelujahs and shouting "Amen! Glory!" &c. It was in vain the landlord exerted himself; before he could get to one room, a louder call from the other end of the building would draw him there, until he gave the matter up in despair, and suffered his obstreperous lodgers to sing themselves to sleep.

Thursday afternoon, rode again to the Camp, saw the most drunken people in the road I ever saw on any other occasion. Many of them, I was told, had families at home destitute, even in this land of plenty, of the common necessaries of life. I could not help groaning in spirit all the way, which was literally perfumed by the odour of the spirit which they had drank. Of course they, the mob, were dreadfully impudent; not to us to be sure—the gentleman who always carried us was uncle to two of the ladies of our party and well known in that part of the country; with him, therefore, we always felt perfectly safe. But he told us he had to dismount, the day before, from his wagon, to rescue some females from insult, two or three times; and that the Methodists had sent for two or three sheriffs to come and keep order.

When we entered the Camp, there was what they called a power-ful preacher, on the stand. He was exhorting the people to repentance with great vehemence and gesticulation. The bad English he used provoked many a smile from his hearers, while another class of his hearers seemed to listen with profound attention, and expressed their approbation by many an exclamation of delight, accompanied with groans and amens. One man fell down near us in strong convulsions; the crowd pressed around him, but the brethren, pushing them back, drew him into a tent, saying he was "full of the spirit," &c. We now got crowded between a woman of most infamous character and some young men, who were holding a whispering dialogue over our shoulders:—astonishing impudence! We removed to another part of the ground as soon as possible; and having regained our escort, proceeded to a bench near the upper end of the Camp. Here, seated beneath some trees, we could look down upon the crowd, though out of its immediate vicinity. Here, too, we could hear most of what the preacher said who was then speaking. One of our acquaintance now advanced from one of the tents, and informed us there had been "quite a riot there" the preceding evening, but that there was "no danger now, as there were several officers on the ground, hired to keep the peace." But there was no solemnity now— all was hubbub and confusion. A sister came up and asked if we intended to stay in the evening, saying they had "such *powerful* meetings in the evening, it was heaven below." I could not but express to her and several others, that I was about tired of it, and should go away with very strong prejudices against Camp Meetings. They assured me it was only because I had not seen enough of them, and that if I should remain with them one evening, they doubted not my prejudices would vanish; and that I should witness such a display of the power of God, as I never saw before; said I was cold, "but she would insure me I should get warmed, if I would only attend their evening prayer meetings." I told her it was not possible for me to stay that evening, as we expected the wagon, to carry us back, before sun-set, and I had engaged to return with my party; but that both they and I intended to pass the last night in the Camp, which would be the next, as we understood they kept up the meetings through the whole of the last night, and we were determined, having heard so much of their evening meetings, to be present one night; and if there was any good to be obtained, to be in the way of it. She appeared to be much pleased, and we separated. Our conveyance now arrived for us and we departed; again past the dark avenue, upon which the shades of twilight were now fast gathering. The brutal intoxication and profanity visible on the road home was truly shocking; and as we went past the stalls, the thought struck me, that these buyers and sellers were after all perhaps the smallest sinners on the ground; that they, who were the means of bringing this tumultuous assemblage together, unless there was some redeeming merit about it that I had not yet discovered, had much to answer for. Dismissing such thoughts however, I resolved not to make a final decision against them, until I had witnessed those meetings upon whose influence they counted so much.

This evening I overheard some ladies, (of whom there was and had been, a respectable number, and a very respectable company at the inn,) teazing their husbands to carry them to the Camp. Their hus-bands positively refused, saying it was not a proper place for females in the evening, and that they could not engage to protect them from insult while passing through the wood and its environs; that they themselves should go but begged the females to remain contented where they were. Children of Eve! I heard them afterwards resolving at all hazards to know where the danger lay, and threatening if their husbands went without them to hire a conveyance and go by them-selves. Whether they carried their point or not I did not ascertain. I saw them all depart from the inn in company together.

A number of young men now repaired to the inn, from the Camp, to get supper, intending to go back again. They appeared in a high frolic, but one of them was taken alarmingly ill. Directly after he was seized with a bleeding at the nose, so violent as to induce the belief that it proceeded from the rupture of a blood vessel. Though at some distance from the apartments of the family, the ladies all volunteered to his assistance. It was a shocking scene, and with the greatest difficulty the effusion of blood was stopped by the variety of applications used. He appeared quite grateful for our kindness, and particularly to the landlord, who immediately after had him carried to a cool room and put into bed; but the effect of the scenes he had just witnessed, had such an effect upon his brain that nothing could keep him silent. As soon as he was comfortably in bed he commenced singing hallelujah, and kept it up for the greater part of the night.

Friday was the last day of the meeting, and I who had now firmly resolved to see it out, and be a judge myself how far it was a work of the Spirit, went prepared to spend the day and night in the Camp. We carried refreshments, and all of our party agreed to keep together; and to ensure our safety, we contrived to go in the suite of an officer of justice, who with his family had stopped at the inn on their way. By the way, we had only occasional glimpses of him after we got there, for being employed by the meeting people to keep order, he was obliged to be on the alert. It was a scene of dreadful confusion to get there in the first place, the road was so full of people, the dust (for the earth had been fairly ploughed up by the multitude of feet) blowing and blinding one. It was a fact that we not half the time could see our horses heads, as we rode on. In the Camp there was great confusion. The crowd had very sensibly augmented. There was a woman exhorting at the stand, and one of our townsmen, who recognized me, and knew I was a great stickler for women's preaching, immediately came up and invited me to go down and hear her. Accordingly we all went down to the stand. A young female whose appearance bespoke her to be under twenty, was exhorting. The first words we distinguished were these, that she "did not want a copper of their money—No I dont want your money," she repeated, "not a copper of your money, only the salvation of your souls," and she exhorted the "young *Ladies*" and the "dear young Gentlemen" to repent, with all the energy she was capable of. Now I who abominate the epithet of *Ladies* and *Gentlemen* in christian exhortations, was turning off, when some one whispered, "Mrs. T—— is going to preach." This lady whom I had once before heard upon a most interesting occasion, was a great favourite with me, and I had inquired several times if she was there. I therefore took my station on a log, and with my companions heard her discourse. The woman speaking was of very mild and pleasing manners—a woman of plain good sense, and exceedingly graceful and winning in her manner, when speaking in a house where her voice could be heard without exertion; but although her discourse which was short, was now, as it always was, good, yet the evidently great exertion she now used, destroyed much of its effect with most of the hearers; the blood looked as though it would burst through her face, the veins of her forehead and temples as well as those of her neck, "swelled up like whip cords," and her mouth, usually of sweet and placid expression, from her efforts to speak loud, was absolutely disfigured.

"Is this the Mrs. T——," whispered one to me, "I have heard you praise so much? Why, I never witnessed such contortions of countenance before." Such remarks proved the woman in my mind to be out of her place, for I had no doubt her discourse was better than any that had been heard there, but the great effort of retaining such a masculine attitude entirely destroyed the effect. She was succeeded by a very bold and uncouth looking young female, whose language was as coarse as her look and manner. She called upon the people loudly to repent "today and save their souls." Some very singular expression she made use of appeared to have an irresistible effect upon a part of her auditors, who laughed aloud; upon which she said she "did'nt care who laughed, she cared for nobody not a snap of her finger," (snapping her fingers in great style.) Another loud laugh. My faith in woman's preaching began to waver, and I was glad to walk off. We observed an African upon a stump at some distance, near the upper part of the camp, collecting a great crowd around him, who were listening with open ears and gaping mouths. Some were wiping their eyes, many shouting, and others grinning. Thither then we bent our course, willing to hear the truth from whatever quarter it might proceed. The first words that met my ear were— "Deble fader of lies; he be liar from beginnin. Some say poor niger hab no shoule. Vel dat I dont know, but dis I know, I got something in my body make me feel tumfortable," (clapping his hands vehemently upon his huge chest). A peal of laughter, long and loud from the profane rabble, was the response. While nothing daunted he continued to go on in the same strain, not in the least interrupted or annoyed by the continued shouts of the mob, who, clapping their hands, kept crying, "go on brother, that's your sort, glory, hallelujah," &c. with all such sort of encouragement. I need not say we did not stay there long; and as no interesting preacher now occupied the stand, we resolved to stroll round and look up some of our friends from the neighboring towns, many of whom we doubted not were there. In passing one of the tents we could not forbear stopping to look at a young woman reclining on the straw in a very languishing attitude, and apparently quite helpless: two or three young men had seated themselves near her and were enquiring how she felt? Upon closely observing her I discovered she was the same young woman whose disordered appearance and extraordinary beauty had struck me so forcibly, and invited so much observation a few day before. It was she, but oh how changed! even in the brief space of time that had intervened since we saw her before. Her bloom was entirely gone, and her haggard look and tangled hair gave her the appearance of something that had recently escaped from a mad house. I shuddered with horror, and thought oh! if you were a sister or daughter of mine how should I feel. Humanity towards the poor victim induced me to draw near and ask her if she had no mother to take care of her? She turned a look of scorn and anger upon me, and then exchanged a look with each of the young men, and they all three laughed, and I walked off convinced I had been mistaken. I afterwards mentioned the case of this young woman to some of the persons on the ground, who undertook to explain to me her situation by telling me she had just come to. Their language I have forgotten, but I understood it to mean that she had gone through a process which they considered as perfecting the work of sanctification, and I afterwards was told by some people at the house, that she was probably the same young woman who had lain two full days in a state of stupor, an unusual long time, and that it was possible her intellects might be affected. Be that as it might, the image of the fair sinner, or saint (for she was no half way character) haunted me for some time. We afterwards looked into another tent where we saw a girl from our own immediate neighborhood, in much the same situation, having just recovered from a state of torpor, and rejoicing with great appearance of happiness. My heart sickened at the sight of her, for I believed her a most accomplished hypocrite, and the end justified my suspicions. In the course of a few months she destroyed the peace effectually of a worthy family, who had taken her from a state of great poverty several years before, and cherished her with all the tenderness of parents. She had previous to this been a Baptist by profession, but after this attached herself to the people through whose ministry she professed to have been recovered from her backslidings, and continued with their society until put out of all society.

Being exceedingly fatigued we were now obliged to give up our plan of remaining in the Camp; the wagon in which we came having arrived with some other persons, we concluded to go home and recruit before the services of the evening. The ride home was no more annoying than when we came; a certain sharp-looking set of fellows seemed to be prowling about the woods, and dodging at every corner—whose very look was sufficient almost to curdle one's blood, but it was now so generally understood that the camp was protected by the officers of justice that none dared to show their colors.

Before it was quite dark we returned, and by the time we arrived, the camp was lighted. I could easily imagine that embellishment added much to the scene. The disposal of the lights which exhibited so many different groups, and displayed the paraphernalia of the tents with such a different aspect from what it appeared in the glare of day, was altogether imposing, or rather witching. For a time we walked, until our protectors returned to the inn to take back the conveyance. We avowed our determination to pass the night in the Camp; the gentlemen remonstrated, urged the fatigue, the exposure to health, the danger, unless we kept close under the wing of some person or persons able to protect us—but all to no purpose, we determined to remain. They promised to return and stay until ten or eleven o'clock, and then they said we must take care of ourselves; and leading us to one of the seats at the upper end of the ground, departed. There were four of us, nevertheless we experienced some little sinking of heart when we saw our protectors depart. From the place where we sat we could see the whole ground; there was a preacher on the stump speaking loudly and vehemently; a black man also on the stand, and nobody attending to either; the noise could not have been exceeded by the confusion of Babel. I could not compose my mind to realize it was a place of worship, although the songs of praise and the voice of exhortation mingled with the groans of despair, and blending in strange confusion with the various dialogues going on, rose each moment on the ear. Prayer meetings had commenced in the different tents, yet there was a continual travelling from place to place—nobody except the immediate actors in the scene seemed stationary for a moment at a time; crowds of people passing and repassing all the time. One woman flew past, throwing her arms abroad, and shouting "there are grapes here and they are good, heavenly times! heavenly times!" A few moments after our ears were assailed with the most piercing shrieks of a female voice, which proceeded from behind one of the neighboring tents. Two of us sprang up and almost involuntarily ran to the place—the other two rather hung back as they afterwards told us from fear, thinking it might be some one murdered, or some terrible assault, a few moments brought us to the spot, and beheld two young women stretched upon the ground, no human creature touching them, screaming with all their strength. Some females from the neighboring tents rushed out to them, and sinking down by their side, began to talk to them all at once. "Sink right into Jesus" said one, "and you will be happy in a minute." I enquired of an old lady standing by what the matter was? she said they were slain, and there was a great many slain there every night. Several persons now raised them to carry them into the tent, and we in a whisper agreed to follow close in the rear, which by keeping hold of each other's clothes and following close upon the heels of those who had borne in the slain, we succeeded in getting into the centre of the tent, where, within a circle formed by the meeting they were laid upon the straw. They, the meeting people, were singing a hymn, which rose to deafening uproar upon our approach. After the hymn, the women commenced praying over them, using many strange expressions and the most violent gesticulation, the power of which was acknowledged by many a groan, shout, and interjection, intermingled with the agonizing shrieks of the slain, which still continued.

The loud Amen, the cries for mercy, the groans of distress, (either real or imaginary) resounded from every quarter, while the triumphant exclamations of those who shouted "I'm full—I'm running over—I'm eating heavenly manna—glory! hallelujah!" &c. &c. were as distinctly heard: and this, this scene of discordant noise and unseemly riot (as it appeared to me) was what they called "the power of God." Forgive, thou insulted Being, the use I am here obliged to make of thy great and dreadful name! Occasionally some of the young men who were within the circle would draw near the young women, whose shrieks gradually changed to groans, and ask, in a low voice, "do you feel any better?" I could not hear that they made any answer. One young man, while the prayer was going on, began to shake violently, and then falling flat upon the straw, exclaimed "God, I'm willing—I will own my Saviour—I will, I will:" at the same time, his feet kicking at such a rate, that the dust from the straw nearly suffocated us all. His feet chancing to lodge, in his fall, just between me and another young lady, we endured no small share of inconvenience. The young lady actually received several smart blows; when a man leaning over our heads (we were seated on a bench) put his cane over and fenced his feet from her, by planting it firmly in the ground.

A few people from our town sat near, and, as I thought, seemed to survey the scene with mournful interest; at least they exhibited none of the animation I have described. "Lord," said one of the women in prayer, "what ails the Providence people?" One young woman uttered a sentence in prayer that seemed to fill the audience with inexpressible delight. It was in allusion to a sentence in the sister's prayer that spoke before, wherein she asked for the crumbs that fell from her master's table. "Give us," said the last one, "not only crumbs, but loaves, good God!" and slapping her hands with great violence. The effect was electric, the Amen was echoed in all the different notes of the gamut, while the expressions of "Come Lord Jesus, come quickly," were heard from different parts of the tent. My soul was momently shocked by those familiar addresses to the Deity, "God, come down here—Jesus come this minute—we want you tonight—we want you now," &c. &c. &c. The din and confusion increased every moment. Stamping, slapping hands, and knocking fists together, formed altogether a scene of confusion that beggars description, and really terrified us. We looked at each other in despair, and then at the door, which was completely wedged up with faces, one above another; no way to get out, and no one to help us; when fortunately the uncle of two of the young ladies, (who had returned to the Camp on foot, after putting up his horses, and who was now standing at the door of the tent,) descried us, and in a moment comprehending our distress, opened a passage to the circle, by saying "a lady faint! a lady faint!" which was echoed by several, either to aid in getting her out, or to increase the confusion, and thus we escaped from the crowd. There was now a general begging among us to return home; but the uncle protested there was no way at present, and we must stay all night where we were. However, as we begged so hard, he despatched a man round the barriers to see if any carriage or wagon could be procured. While search was making, he advised us to walk around the ground; as hundreds, probably thousands, were then doing, thinking we should be safer to be moving with the crowd, than to sit down any where outside the tents. As we passed one of the tents, where the confusion could only be equalled by the one we had left, we distinguished in prayer that remarkable sentence, "the Lord is in his holy temple: let all the earth keep silence before him."[9] What a place for its repetition!! One young man began to pray, who got so animated that he kept asking to die; exclaiming, "Lord, I want to die. I'm ready to die and fit to die, and Lord, I want to die to-night." Loud shouting and clapping of hands followed.

We now passed a tent entirely closed, fastened down, and dreadful groans within: they appeared to proceed from one voice, and that of a woman, and evidently betokened great bodily distress. One of the gentlemen just behind us said he was determined to see what was the distress, and began unfastening the curtains: we had been forbidden to raise it, by a brother who stood outside, but after the young man had got it part way up, a minister from within called out "come in and see the power of God." Thus invited we entered, and behold, a young woman laying flat upon the straw, in great apparent agony, calling in frantic terms for the coming of the "Holy Ghost." I saw no other inmate of the room, except the minister just mentioned, but

---

9. Hab. 2:20.

upon our coming out, several methodists passed in, and we heard them a moment after singing around this distressed creature, "Die in the arms of Jesus, Die in the arms of Jesus &c." That woman had every appearance of being in strong hysterics.

We had just met with a party of friends from the village of C——— who learning our distress, kindly offered a seat for me to return with them to the inn, where they were to pass. It was now eleven o'clock, and my companions consented I should leave them upon the promise that I would not rest until I had found some wagon or carriage of some kind to come after them: while passing to the barrier where the wagons were stationed, we passed a tent where a young female, apparently quite gone, was supported in the arms of a worthless fellow, who had lately gone from our neighbourhood, no one knew whither: they were just without the tent door, and he was trying to bear her in. A fear for the safety of the girl induced me to ask some one near to rescue her, which they attempted, when out burst two or three men to the relief of their brother, as they called him, and forbade any interference.*

We felt rejoiced that this was the last night of the meeting, for the camp began to smell very offensive. Many were remarking that the

---

* I learned afterwards that this young man went at the opening of the camp, with a small bundle, into one of the tents, and there continued through the whole meetings. This singular character was once a regular member of a church (not Methodist) in P. and was thought too simple to be set aside, though known to be a most inveterate liar. On one occasion, while travelling through the country, he passed himself off as the grandson of a venerable clergyman of the Episcopal Church, well known in that vicinity, and was entertained at the tables of some of the most respectable people in the town—invited to read prayers in the church, as he professed to be a candidate for the ministry. This he did, and escaped undiscovered. He then proceeded to one of our western settlements, where, chancing to hear some inquiry made respecting a good old lady in his town, lately deceased, he told them she had left an immense fortune, which she amassed by keeping a hotel. This family, who chanced to be heirs at law, entertained him with cordial welcome, and actually came a journey of many hundred miles to claim the property, and found the whole a falsehood. The author of that story, and of many other similar deceptions, is among the Shaking Quakers in New-Lebanon, where he has at last, as he says, "made his principles bend to his temporal interests."

danger to health would be very great should the meetings continue twenty-four hours longer. The people who conveyed me out, got along very well, through a road used for a cart path, and which appeared much more safe and quiet than the great entrance. We saw a good deal of dodging about, though upon comparing notes with others I discovered I was not solitary in hearing and seeing strange things. One lady who had been invited to drink tea in one of the tents, observed she had been much shocked by a man coming in and inviting her to stay the evening. He went in shaking violently and saying "we shall have the Holy Ghost here to-night!" and said a little niece of hers who stood by, "do stay aunt Polly, for I want to see him." These kind of anecdotes were long rehearsed, but I met with no solitary being who appeared to have got any good by going. Heaven grant there might have been some. With a great deal of difficulty I persuaded a man to go down with a carriage for the other ladies. At last, after finding two men to go and assist him, he went. He said there had been one carriage just before which had all the harness cut off of it at the entrance of the wood. I could not rest until the whole company were safely housed. They returned about one o'clock in the morning. The inhabitants of the neighbourhood long had cause to remember that meeting. The effects of it were distinctly visible. Fences torn to pieces, and fields of grain wantonly trod down and destroyed, with other excesses, absurd and unnecessary, bear witness to the little reformation in morals the meeting had occasioned; but over and above all, the haggard and jaded looks of those people, when they commenced their homeward march on the following day. A rain, the first the earth had been blest with for some time, fell on that day, and many of them must have been caught without a shelter—some with little infants in their arms. One I saw at the camp which the mother told me was three weeks old!

It must be obvious to every person, of common sense, that if camp meetings exhibit such scenes to moral persons, to those who penetrate the recesses in their neighbourhood the view must be still more revolting. Stories have been told and still are, that almost stagger credulity itself, and they carry with them this proof of their authenticity, that the most depraved and abandoned of the human species, are always fond of resorting to them. If the writer of this true sketch can be a means of opening the eyes of any well disposed persons, who have hitherto been disposed to uphold them, it will be a source of lasting satisfaction, and a full reward for all the resentment which ignorance and fanaticism may award.

Observations on the foregoing Narrative

Upon looking over the preceding pages the author has not been able to discover any mistakes, though there are many things which may be liable to misinterpretation, and some things omitted which the limits of the book would not permit her to discusss. Of the first of these, the reflection upon spreading the report in Providence, which proved so disastrous in the after life of Miss Cornell, is not meant to be attributed to the merchants spoken of—the scandal we know was transmitted to the public through other organs. And with respect to the letter from Bristol to one of the witnesses, containing three dollars, and which is said to be the sum actually due her, dating from the time she was summoned, it is due that witness to state, that in her narration to the author of this, she did not say it was not due her, because she was totally ignorant on that head, but she expressed some surprise that they should have "left it until after Avery was taken again"—particularly as no recompense had been tendered her for her attendance at Newport where she had been "summoned by the prisoner and detained much longer."

There is one subject upon which we wished largely to have descanted in this work, but upon which a few words must suffice: that is, the great injury and injustice which the publication of the life and character of Sarah M. Cornell, has done to that class of young women whose lot in life has compelled them to labor in a manufactory. Many have taken the liberty to say that if all those disgusting particular were true, it proved to demonstration that "vice was not regarded among that portion of society as it was in any other community; that there was little regard to morals among them, or that persons could not have been tolerated and associated with as we know she was—and finally that it ought to be a warning to parents not to let a daughter go to those places, which was going to certain ruin." Now nothing can be more unjust than this. There is no person who deprecates the practice of sending little children into a cotton manufactory more than the author; she avers with truth that she has often been affected to tears at the sight of the little innocents, compelled to leave their beds before the rising of the sun and labor until long after its going down in those establishments, and that perhaps to support some idle, drunken father, or miserable, unfeeling mother; but when she has again seen healthy, sprightly and well educated girls, laboring to assist some widowed mother, or to give education to some half dozen little brothers and sisters, her feelings have received a different impulse. There is no way that grown up girls in the present state of society can get better wages— nor where their payment is so sure. And the privilege of working in manufactories to such is a great one. That these girls are careless of their conduct or their company is scarce ever the case—and the author has known numbers despised and shunned, and hunted from the manufacturing villages, upon a charge of a much less serious nature than any of those brought against S. M. Cornell, that is, where they had no meeting to shelter them—where back-slidings and recoveries, expulsion and reinstation, were a common thing. In such a case perhaps it might not be known out of the meeting. Why, if it were publicly known, as it ought to be, a girl guilty of half the offences she is charged with, in the state of Massachusetts or Connecticut, would at once find herself in the House of Correction.

The publication of this matter however has had one good tendency which is obvious. It has generated a suspicion of those noisy, ranting professors, who go about interrogating every one they meet, to know "if they love the Lord? if they enjoy religion? if they are not ashamed of Jesus?" &c. &c. which none but grossly ignorant or hypocritical people ever think of asking. We hope and trust it has not lessened the respect felt for those modest, practical and retiring christians, who mind their own concerns, and pursue the even tenor of their way, without seeking to obtrude themselves or their religion, except where propriety sanctions, and principle and duty authorize them to do it; and these occasions are not rare. There are daily and hourly opportunities for the real christian to shew forth the beauties of holiness, without disgusting people with impertinent interrogations, discovering an impudence and boldness inconsistent with their sex and professions.

Since writing this book we observe there has been a great hue and cry among a certain class—that religion was in danger from dwelling upon this subject—that it was better to have it smothered, or in their language, *dropt,* and that every christian who lent his aid to keep it in memory, was strengthening the hands of infidelity. To such we would say, we view the subject in a very different light, and we consider it as a very suspicious circumstance in professors or ministers of the gospel when they wish to smother "spiritual wickedness in high places." We firmly believe that religion is not so inseparably connected with E. K. Avery, so identified with him, that it must rise or fall with him, or indeed with any other preacher. We have always believed that the existence of counterfeits, was itself a proof there was real coin somewhere, and have been accustomed to consider the Christian Church as a net cast into the sea which gathered fish of every kind, both bad and good. Our Bibles tell us "there will be deceivers in the last days."[10] We consider the scripture as fulfilling, and that these enormities being foretold and now accomplishing, proves them true; but we are not warned to spare such offenders

---

10. Possibly a paraphrase of 2 Tim. 3:1–5.

because of their professions, but on the contrary, "that judgment must first begin at the house of God."[11] And we believe whoever is able to assist in this and in pointing out the difference between true and false religion, is doing society and religion itself a great service; and though men may mistake our motives, we can appeal to the Searcher of all hearts for the purity of them, and we look with hope and confidence for his approbation at the resurrection of the just.

---

11. 1 Pet. 4:17.